JOHN TAY

mining entrepreneur

1779-186

John Taylor in 1825. From a portrait by Sir Thomas Lawrence for the Athenaeum Club.

John Taylor

mining entrepreneur and engineer

1779 — 1863

by

Roger Burt

MOORLAND PUBLISHING COMPANY

ISBN: 0 903485 36 2

Printed in Great Britain by
Wood Mitchell & Co Ltd, Stoke-on-Trent

For the Publishers
Moorland Publishing Company
The Market Place, Hartington,
Buxton, Derbys, SK17 0AL

Contents

Illustrations

Introduction

Until the end of the eighteenth century, most of the major mining districts in England and Wales were almost entirely self-sufficient in capital, entrepreneurship, expertise and labour. Where any of these factors were locally deficient, they were largely drawn in from surrounding areas and there was relatively little movement between mining districts. By the early nineteenth century, however, this isolationism was rapidly breaking down and by the mid-century a truly national industry had emerged, drawing on a national capital market with large-scale inter-regional migration of managerial skill, technique and machinery. These changes were produced by rapid technological development in the industry and escalating capital costs from the need to exploit mineral lodes at greater depths. This new era had many pioneers, largely drawn from the South-West, then the most advanced sector of British mining, but the most outstanding of them was John Taylor. It was he more than anyone else who was responsible for the innovation and national diffusion of the latest Cornish technology and he that was the first to mobilise the immense capital resources of London and introduce them into the mining districts.

When I first came to this subject, Taylor appeared as an extraordinarily shadowy figure. His name constantly cropped up in conjunction with large-scale projects in every major mining district, but there was little or no sustained biographical material to tie these activities together. The problem was compounded by the succession of *John* Taylors who were connected with mining and whose activities many authors mistakenly attributed to this one man. This study is still far from providing a complete document of his life and work. It does little more than pull together scraps of information and accounts of particular activities gleaned from a multiplicity of different sources and present them in a general chronological and topical form. Unfortunately, it is not possible to go much further. The extensive records accumulated by the family business which John founded were tragically destroyed by bombing during World War II and one is obliged to rely on old surviving out-letters scattered through the libraries of the great London societies, the British Museum and Public Record Office, and local record offices throughout the country. It will have to wait for the rediscovery of further source material and the detailed research of other writers before the story can be substantially extended. This is particularly true for the story of John Taylor's private and family life. What I have presented here is essentially a business biography and a record of his public interests and contributions. I have found little of his private correspondence and virtually nothing relating to the character and style of his personal and domestic affairs. One is tempted to speculate on these matters, but my own suggestions would be no more valid than those of the careful reader.

Acknowledgements

Of the many people who have provided me with information for this volume, I would like to offer my particular thanks to Lynn Willies and Colin Short who made available their own notes and research on aspects of John Taylor's life and career. I am also very indebted to Christopher Schmitz of the University of Bath who conducted much of the research on Taylor's family background and the early stages of his career and to Professor R W Randall, whose work on the Real del Monte mining company, provided the basis for Chapter 3. The Norfolk Record Office, the Clwyd Record Office, the Science Museum, C. J. Williams, R. H. Bird and J. A. Robey have kindly provided illustrations and valuable assistance for the presentation of the book.

CHAPTER 1

Family, Childhood and the Foundations of a Reputation

John Taylor's origins were inauspicious for a man who was to be acclaimed at the end of his life as the 'patriarch of British mining'. Unlike the other great mining engineers and entrepreneurs of the eighteenth and nineteenth centuries he was not a native of the mining districts where he was to make his reputation. He was born, on the 22 August 1779, in 75 Gildengate, (now St George's) Colegate, Norwich,[1] in the heart of agricultural East Anglia and about as far from any form of mining activity as it is possible to be in England.

John's family, who were of relatively modest means, had been established in Norwich in 1733 when his great grandfather, a Unitarian minister, had moved there from Scotforth in Lancashire. The first of several generations of *John* Taylors, he was largely responsible for the foundation of the Octagon Chapel, still a well-known landmark in that city and through his hymn writing and other religious tracts earned an enduring reputation as a major figure in the development of the Unitarian church. His only son, Richard (1719-62), established himself as a successful wool merchant in the city and in 1748 confirmed a social position of some influence by marrying Margaret Meadows, the daughter of a former mayor and herself related (through marriage) to the powerful Martineau family of Norwich. Later in life, however, Richard appears to have lost the greater part of his fortune, reputedly as a result of the Lisbon earthquake of 1755, and to have died at the relatively early age of 43, leaving his widow and several young children to fend for themselves in greatly reduced circumstances.[2]

Richard's eldest son John, only 12 years old when his father died, spent his early years apprenticed to a cloth manufacturer in Norwich and later moved to London where he acted as a clerk to the bankers Dinsdale, Archer and Ryde. He appears to have had some success here and returned to Norwich in 1773 where, together with his brother Richard, he established a yarn manufacturing business. Although this never produced a great fortune, it evidently did reasonably well and within a few years John had regained his father's previous position in the community.

John became active in municipal and social affairs, taking an active part in

1 and 2 The Octagon Chapel. Norwich; *(top)* the exterior in 1828, *(bottom)* the interior in 1848. *(Norfolk Record Office)*.

the foundation of Norwich Public Library in 1784, and continued the family tradition as a prominent lay member of the Octagon Chapel and hymn writer. Politically he was a radical and reformer, and together with his cousins, the Martineau family, he emerged as a leader of the Whig party in Norwich. These activities brought him into frequent social contact with the aristocratic élite of the district, including the royal Duke of Sussex, the Earl of Albermarle and the prominent agricultural reformer, Mr Coke, the Squire of Holkham. John's wife, Susannah, who he married in 1777, shared equally in her husband's social and intellectual interests and added a moderating influence, so essential in those days of high party feeling and religious prejudice, which enabled their home to become a meeting place for many of the foremost literary figures of the day.[3] Of these gatherings, Janet Ross was later to write 'In the old fashioned parlour of St. George's, Colegate, might be heard the most brilliant conversation, as eminent people of every opinion gathered round the hospitable, unpretentious fireside'; their number including Sir James MacIntosh, Sir James Smith, Crabbe Robinson, Dr Southey and Mrs Windham.[4]

It was into this moderately prosperous, intellectually active and staunch dissenting family that our John Taylor was born, the first of seven children.[5] The youngest of these children, Sarah, was later to write proudly that her mother never buried a child; a great achievement in an age of high infant mortality. The family atmosphere and their parents' values made an indelible impression on all of their children. Throughout their lives they remained a close-knit family with frequent large gatherings in their Norwich home. All of their sons became prominent members of many of the most influential learned societies of the day and all conformed to the high moral principles of their faith.

None of these characteristics is more clearly illustrated than in the career of their son John. Later to become a Fellow of the Royal Society and a founder member of the British Association for the Advancement of Science and of University College London, he emerged as one of the most respected and influential men of science of his day. His mother, cut off from the imposing chambers of the London Societies and reputedly obliged to conduct her own disputations with Brougham, Southey and MacIntosh while darning her boy's grey worsted socks,[6] might well have been impressed by her son's achievements. She undoubtedly also would have been proud of his almost legendary reputation for honesty, trustworthiness and propriety in all his dealings. Susannah Taylor and her husband drilled all of their children from an early age in the strict moral code which distinguished many dissenters during the period. One particular aspect of this code – the meticulous control of financial matters and the careful

3 The Taylor family tree.

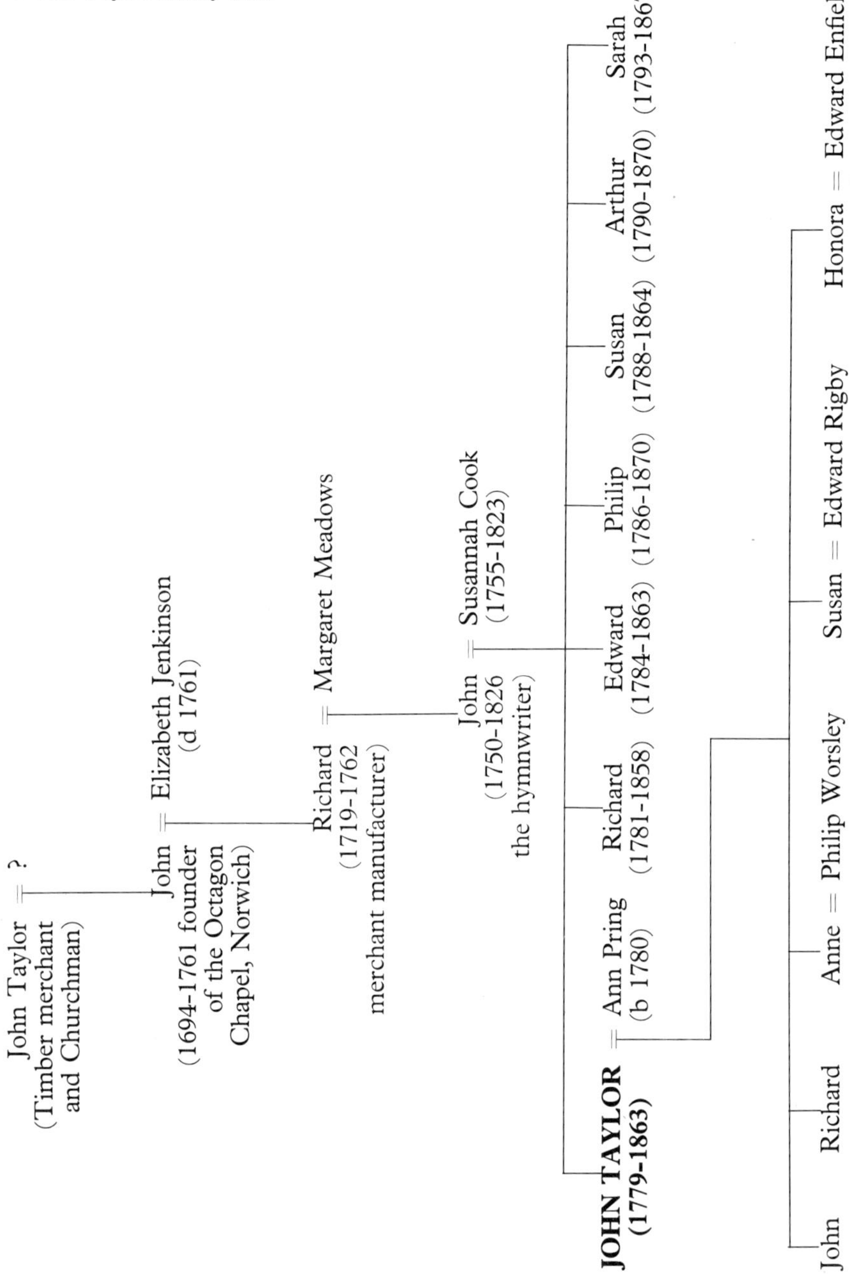

4 John Taylor Senior (1750-1826), painted by N. Branwhite and engraved by R. M. Meadows. *(Methodist Magazine,* 1808*)*.

5 *top*, Susannah Taylor (1755-1823); *centre left*, Sarah Taylor, John's sister, later Mrs Austin; *centre right*, her grand-daughter, Lady Duffgordon; *bottom*, her great grand-daughter, Janet Ross. (*The Critic* [New York], 1893).

avoidance of debt – must surely have been emphasised by the haunting memory of Richard Taylor's business failure and the consequent perilous problems which had beset John's father's early years. The lessons were certainly not lost on John. His future career, contribution and reputation was founded on a proven unflinching honesty in an industry which was notorious for its rogues and deceivers, coupled with a financial genius derived from the relentless pursuit of just and proper economy. It was these qualities, rather than any great scientific or technological originality, which were to underpin his success and make him the most popular and sought-after financial adviser and treasurer of his day.[7]

The influence of Susannah Taylor was not confined to the social and moral development of her children; she also played an active role in their formal education. She provided the initial instruction for all of her children in the rudiments of arithmetic, grammar, and foreign languages and for her eldest son, John, in whose education she is said to have taken a special interest, she provided mathematical instruments and a turning lathe, 'to encourage his boyish taste for mechanical pursuits'.[8] When her boys were old enough, this private tutoring was furthered by sending them to a local day school, founded in 1787. It was run by Dr John Houghton , the retired minister of Wem in Shropshire, in conjunction with his son, the Rev Pendlebury Houghton, who took up the post of joint pastor of the Octagon Chapel.[9] The Houghtons apparently had a reputation as strict disciplinarians but provided what was considered a liberal and advanced curriculum for their students, grounding them in chemistry and other

6 and 7 Two of John Taylor's brothers; *(left)* Richard (1781-1858) *(National Portrait Gallery)*, original with the Linnean Society), *(right)* Edward (1784-1863), Sheriff of Norwich and Gresham Professor of Music *(Norfolk Record Office)*.

scientific subjects as well as the more usual classical studies.[10] The education of John and Susannah's sons was completed by apprenticing them to some suitable profession. Unable to offer their children a secure financial future or to set them up in their own businesses, they endeavoured to ensure that they had the wherewithal to make their own way in the world. Thus John was apprenticed as a land surveyor and civil engineer, Richard as a printer, Edward as an ironmonger, Philip as a surgeon and later as a druggist, and Arthur also as a printer with his elder brother, Richard. It is notable that none of them was put to their father's own trade as cloth and yarn merchant and that with the single exception of Richard, all made their main mark outside their original professions; John in mining, Edward in music, Philip in engineering and Arthur in archaeology.

It was at the end of his apprenticeship that John Taylor's career took a sudden and unprecedented turn. Sometime in 1798, during his nineteenth year, he paid a visit to mines in the Tavistock area of Devon with members of the Martineau family. Like the Taylors, they were strong Unitarians and there were longstanding family friendships and ties through marriage. The Martineaus appear to have been more prosperous than the Taylors, receiving a large income from interests in brewing, banking and sugar refining, and to have acquired shares in Wheal Friendship, one of Devon's largest copper mines near Mary Tavy. Supposedly impressed by John's comments on the mine, and possibly intent on finding a post for the eldest son of their close friends, the Martineaus took the almost unprecedented step of initiating an invitation to John Taylor to take on the management of the mine.

This was a strange decision to say the least. In an industry where mine managers were invariably appointed from the ranks of those with long associations with mining, either as workers or investors, to bring in such a young and almost totally inexperienced outsider must have seemed wildly impractical and a staggering example of nepotistic influence. It is possible that these years of high war-time prices and a boom in mining activity had created a temporary shortage of good managerial talent but few other mines appear to have made such apparently rash appointments. It could also have been that John's appointment was made with a specific view to employing his newly gained talents as a surveyor and civil engineer. At this time the main problems confronting Wheal Friendship were not those of production but of the efficient dressing of the ore and its carriage from the mine to the nearest point of easy transhipment at Morwellham, at the navigable limit of the Tamar. The improvement of dressing operations through the more intensive use of water power and the planning of some new transport route, such as the Morwellham Canal undertaken some years later, might already have been at the back of their minds. However, this need not necessarily have entailed giving him control of other aspect of the workings. Even in this age of 'amateurs', when great achievement and responsibility often came early in life, John Taylor must have appeared extremely young and inexperienced for such a demanding appointment.

Whatever the reasons behind his appointment, once in office John proved an immediate success, more than justifying the faith of his supporters and his annual salary of £100.[11] Within a year he had developed sufficient confidence in his own judgment and abilities to take a direct financial interest in the re-opening of the neighbouring Wheal Crowndale copper mine and the success of that venture, for which he also acted as agent, helped consolidate his rapidly-rising reputation.[12] By 1801, when he was still only 21 years of age, he had initiated the construction of the complex system of leats which were to make Wheal Friendship's dressing floors the most fully-mechanised water-powered installation in the South-West and the mine itself one of the most profitable copper producers in the neighbourhood. The principle embodied in this project, of increasing or soundly founding the long-term profitability of a mining venture on heavy investment in mechanisation and the innovation of the latest technology, was one which Taylor was to cling to throughout his business life. It was also one which distinguished him from the general run of mine managers and speculators who usually endeavoured to minimise investment outlays and look only for a quick profit.

In 1801 John was joined by his younger brother Philip at his new opulent home, Holwell House, Whitchurch, near Tavistock, the former seat of the Glanville family.[13] Now 15 years old, Philip was apprenticed to study surgery and medicine under Dr Harriers, a relation of his mother with a practice in Tavistock. However, Philip appears to have been little taken with blood letting and the other unpleasantries of the medical profession of that day and to have given most of his time and interest to the mineral and engineering work of his brother and his new found friend, Arthur Woolf. This state of affairs continued until John's marriage in 1805 to Ann Pring of Awliscombe, near Honiton.[14] Philip returned to Norwich where he joined Dr Fitch in a large chemist and drug business and set up a factory for making pill boxes with machinery.[15] Meanwhile John and his wife settled to bringing up their own family, the eldest son John being born in 1808 and their second son Richard in 1810.

8 and 9 The Tavistock Canal, with *(bottom)* the east end of the Morwellham Tunnel.

Almost as soon as he was established at Friendship and Crowndale Taylor began to give attention to the serious problems of transport and distribution for the ore which his mines produced. The local terrain and poor roads made wheeled traffic almost impossible and the need to send all the goods brought to and from the mines by packhorses – at rates of up to 12s per ton mile – so increased overall costs as to more than wipe out the savings made by improvements in mining and dressing operations. As early as 1800 he had formulated a plan to solve these problems by constructing a water link between the Tavy and the Tamar. This plan was undoubtedly inspired by the contemporary mania for canal construction as well as by the success of the technically similar leat systems which he was constructing at both mines. In the detailed planning of the canal he was heavily influenced by the technical writings of John Smeaton[16] and, with an eye to every possible economy and profit, developed the project to take account of its potential for mineral exploration. When the final plans for a 4½ mile canal link between the centre of Tavistock and Morwellham Quay were put to Parliament at the beginning of 1803, it was clearly understood that the situation and direction of the necessary tunnel were chosen, 'for the express purpose of cutting all the tin and copper lodes in Morwelldown' and to 'work them in the most profitable way'.[17] Although the total cost of the project must undoubtedly have been increased by such planning, it was hoped to recoup at least part and possibly the whole of these costs from working the new deposits which were expected to be encountered along the route.

Early in 1803 an Act was obtained which authorized the construction of the canal and the formation of a company with a capital of £40,000 in £100 shares.[18] Most of the capital was subscribed locally, though members of the Martineau family took up at least 65 shares and the Birmingham Copper Company, whose interests were vested in Taylor's hands, took 50. Taylor himself, who was appointed engineer in charge at the company's first meeting, in Tavistock on 16 March, appears to have taken no personal interest until 1817, when he purchased shares which he retained throughout his active life.[19] Work started towards the end of August 1803 and at first went well. As expected, the copper lode which had been profitably worked at Wheal Crowndale and the Gunnis Lake mines to the east and west of the canal was almost immediately cut by those driving the tunnel into the northern side of Morwellham. A new venture, named Wheal Crebor, was soon underway and things seemed set well for the future.

However, the good fortune was short-lived. Tunnelling in the hard slate rock proved more difficult than anticipated, being complicated by serious water and ventilation problems and the workings at Wheal Crebor, though producing an increasing amount of ore, eluded profitability until 1813. To expedite the project, Taylor recommended the sinking of more shafts into Morweldown from which sections of the tunnel could be driven and installed new water-powered pumping engines to cope with the serious flooding problems. In 1808 he also designed and installed new ventilation machinery, operating on the principle of combined evacuation and induction. Two years later he was awarded a medal by the Society of Arts for this machinery, details of which were published in their *Transactions* of that year. The completion of the surface canal from Tavistock Quay to the nothern end of the tunnel by the end of 1808 alleviated some of the costs of construction by cheapening the cost of haulage of materials to and from the workings, but it was not until August 1816, after 13 years of driving, that the project was finally completed.[20]

During these years, Taylor's attention was not entirely taken up by the construction of the canal. He continued as manager of Wheal Friendship and Wheal Crowndale and continued their proper planning and development. The workings at Wheal Crebor were also under his direct supervision and in 1807 he initiated a project for a long inclined haulage plane and engine which set the seal on the profitability of that working when it came into use in 1811. So successful was it, that a similar device was later constructed to lower goods from the southern end of the tunnel down to Morwellham Quay[21] and the construction of an incline became almost a trade mark at many Taylor mines. By 1808, after less than ten years in the South-West, John had emerged as the leader and spokesman for the local mining interest. In that year, for example, he was appointed to represent nine local mines in negotiations with the Duke of Bedford to secure a reduction in mineral royalties, consequent on the disastrous wartime fall in copper prices and increase in mining costs. Arguing persuasively that many mines had become unprofitable and that the future exploitation of the Duke's mineral property was seriously threatened, he succeeded in obtaining a reduction in duty, from one eighth to one tenth of total production, from January 1809.[22]

REFERENCES

1 My thanks to the Reverend J. R. Burton for this information.
2 P. M. Taylor, *A Memoir of the Family of Taylor of Norwich* (1866), p2.
3 *Ibid*, pp9-13. See also S. Smiles, *The Lives of Boulton and Watt* (1865), p367.
4 J. Ross, *Three Generations of English Women*, (1888), **III.**

5 They were John (1779-1863); Richard (1784-1858); Edward (1784-1863); Philip (1786-1870); Susan (1788-1864); Arthur (1790-1870) and Sarah (1793-1867). See National Library of Wales, MS 11138D, Letter from Sarah Taylor (otherwise Mrs. Austin).

6 Ross, *English Women.*

7 John Taylor's obituary in the *Mining and Smelting Magazine,* May 1863, said of him, 'Although never strictly a scientific man, Taylor occupied for many years a prominent position in the scientific world.'

8 P. M. Taylor, *Memoir,* p15.

9 J. Taylor and Edward Taylor, *The History of the Octagon Chapel, Norwich* (1848), p54.

10 A. T. Gage, *A History of the Linnean society of London* (1938), p25.

11 This was about average for the captain of a small mine at that time though Taylor apparently considered it 'an uncommonly handsome allowance'. See F. Booker, *Industrial Archaeology of the Tamar Valley* (Newton Abbot, 1967), p103.

12 Devon Record Office, L 1258, No 84.

13 Lyson and Lyson, *Devon* (1822) **II,** pp555, 615.

14 D. P. Alford, *The Abbots of Tavistock* (1891), p319. Ann Pring, a year younger than John, was the third daughter of Daniel Pring and Honour Rowe and had been brought up by her uncle Thomas Burnaford. Her elder brother, Daniel, was a sea captain in the Royal Navy.

15 P. M. Taylor, *Memoir,* p44.

16 *MJ*, 11 April 1863, p251.

17 DRO, Bedford Collection, Canal Papers No 21.

18 The story of the construction and operation of the canal has been eloquently told by Frank Booker in his *Industrial Archaeology of the Tamar Valley.* To minimise repetition attention here will be confined to John Taylor's role in the project and his particular contributions.

19 In view of the poor performance of the canal during the 1820s and after, this continued interest suggests an unusual nostalgia and attachment by Taylor to his early achievements. This was perhaps also reflected by his life-long interest in the affairs of Wheal Friendship, though the continued profitability of this venture provided a more sound economic rationale for his behaviour.

20 DRO, Bedford Collection: Tavistock Canal Papers.

21 Booker, *Tamar Valley,* p113.

22 DRO, L 1258, Letters 83.

ABBREVIATIONS

BM British Museum.
DRO Devon Record Office.
PRO Public Record Office.
UCL University College of London.
BPP British Parliamentary Papers.
DNB Dictionary of National Biography.
MJ Mining Journal.

CHAPTER 2

London and Diversification

I An Interlude in Manufacturing

In the spring of 1812, when already established as one of the leading practical experts in West Devon mining and with the full fruits of his work still to be harvested, John Taylor attempted what at first sight appears to have been a sharp about-turn in his career. He resigned all of his active commitments in the area and left to join his brother Philip in establishing a chemical works at Stratford in Essex. On leaving Tavistock he received a tumultuous farewell from hundreds of miners and their families whose respect and regard he had earned for his fairness in the running of the everyday affairs of the mines. The reasons for this sudden move are obscure. It could be that at still only 32 years of age he simply felt that he had already achieved all that he could in this corner of England and sought the challenge of new problems: particularly the taxing scientific and technical problems of the still infant but rapidly developing chemical industry. From the course of future events, however, it could equally well be argued that he planned indirectly to expand and diversify his mining interests by moving from the provinces to London. He might have already astutely realised that London was fast becoming the centre of national mining finance and might have hoped that from there he could extend his interests into other mining areas and penetrate the higher offices of company management.

At first, John's interests were almost entirely absorbed in the new chemical company. The two brothers, who were associated with various other investors, including the Martineaus and possibly the Ricardos, organised a broad division of function and responsibility between themselves, with John concentrating on metallurgical chemistry and Philip on the design and installation of apparatus. For both of them it was a time of abounding ideas and experimentation, kept in check only by the more businesslike caution of their partners.[1] At first they concentrated on the manufacture of vitriol, using imported pyrites and brimstone, but they soon diversified in a separate company, into the manufacture of apparatus for producing gas from oil.[2] Philip appears to have been mainly responsible for developing this process,[3] though its patent (No 3929) issued on 14 June 1815, was in John's name. Just eight days later, they received another patent

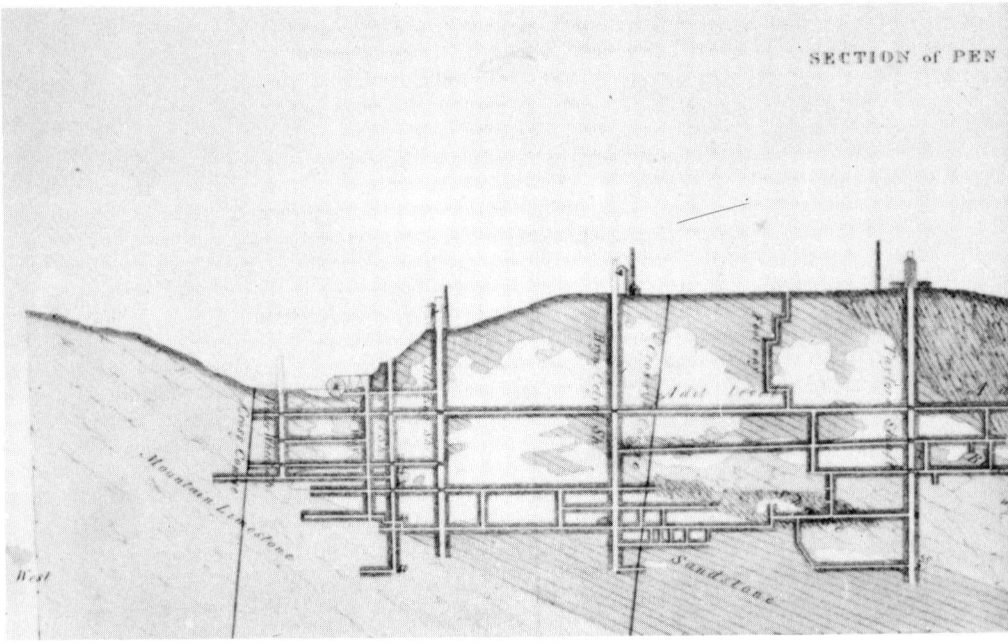

(No 3933) for a process of refining and purifying sugar using high pressure steam[4] and by 1817 were firmly established in that business as well as being involved in other experiments with high pressure boilers and steam engines.[5] In 1819 the brothers were joined by a third partner, John Martineau, a son of the family which had given John his first start at Wheal Friendship. The following year they expanded their business yet again, this time into mechanical engineering, manufacturing portable printing machines for Mark Brunel. At this point, however, increasing reinvolvement in mining affairs began to make heavy demands on John's time and he was obliged to withdraw from further formal involvement in the business, which continued under the name of Taylor and Martineau. Nevertheless, he retained a keen interest in the business and introductions to his brother's business and scientific colleagues greatly expanded and diversified the range of his acquaintainces.

II Halkyn

John Taylor first renewed his involvement with mining in 1813, just one year after leaving Tavistock. At first, however, this new interest was not in

10 Section of Bryncelyn (Pen y Fron) Mine, from John Taylor' report on the Mold Mines, 1827. *(Clwyd Record Office).*

the South-West, the scene of his earlier successes, but in Flintshire, as agent to Lord Grosvenor, the owner of rich lead mines around Halkyn mountain. Taylor might have been assisted in obtaining this post by William Adams, the Duke of Bedford's agent, who had developed a high regard for John's abilities when still in Tavistock, and it is even possible that the promise of such an appointment could have been a deciding factor in his departure for London. Whatever the background, the appointment marked a new stage in Taylor's mining career and one of which he was determined to make a success.

Although the conditions at Halkyn were considerably different from those at Tavistock, Taylor's firm grasp of mining technology and economics meant that this success was not long in coming. He introduced the latest Cornish pumping and dressing machinery to an area which had seen little change for many decades and appointed a Cornish mine captain, Absolem Moore, to ensure efficient day-to-day organisation. Although this sudden influx of new techniques and strange managers was not greeted with enthusiasm by the local miners, many of whom rioted against the new terms of employment introduced in 1822, Halkyn was soon the most profitable mine in the district.[6] In 1829, for his success in assisting Flintshire mines generally to withstand a serious price depression, he was honoured in a celebration dinner at Holywell, by a group of local smelters and mine owners. Lord Grosvenor, who owned extensive mineral property, much of which he worked on his own account, could well be pleased with Taylor's performance. It undoubtedly did much to secure John other similar posts and started him on a process of geographical expansion which took him into every major metal mining district in the British Isles. The experience and expertise which he acquired in these posts also contributed immensely to the commercial success of his own ventures which he later promoted at home and abroad.

III Return to Devon

At the same time as he was moving into new areas and aspects of mining activity, Taylor also gradually reinvolved himself in the Tavistock area. The first steps were taken in 1815 when he became involved in negotiations between the local mining adventurers and the Duke of Bedford for a further reduction in royalties, from one tenth to one twelth, because of the continued post-war depression in mining. The success of these negotiations and the record of Taylor's earlier achievements in the district still fresh in mind, led the mining companies to try and tempt him from his self-imposed exile as a potential saviour during these difficult years. The Tavistock Canal Company had flattered Taylor in January 1815 by voting him 100 guineas (£105) in appreciation of his former services and in November 1816 the Wheal Friendship company, at their annual meeting, in Baker's Coffee House, Charge Alley, Cornhill, resolved that,

> from the long residence of Mr. John Taylor in the neighbourhood of Tavistock, the very judicious and able manner in which he managed the mining interest and other works in that vicinity and the great interest he has continued to evince in their prosperity since he has given up their superintendence, render it highly desirable that the Proprietors should avail themselves of the services of that Gentleman under the present unprecedented difficulties.

Taylor was invited to accept the office of Secretary to the Committee of Management with wide powers of control over finance and day-to-day working. He took up the post for a remuneration of £50 per annum, a sum which was increased the following year to 100 guineas per annum with a gratuity of another 100 guineas, for past services.

During 1817 and 1818 John's handling of the mine's affairs was severely criticised by a certain Mr Burge, a member of the management committee, but at the annual general meeting in December 1818 he was given an overwhelming vote of confidence by the shareholders, a further testament of 100 guineas, and was confirmed as manager and treasurer of the whole London department, 'with full power to direct the ordinary operations' of both Wheal Friendship and the neighbouring Wheal Betsy.[7] Burge withdrew from the committee and John's voice continued as a dominent force in the affairs of the company for decades to come.[8] In his future, as well as his past policies, he could no doubt count on the support of the Birmingham Mining and Copper Company and the Martineaus, who both held interests in the company, and from the early 1830s he was joined as a shareholder in the company by his sons, Richard and John.[9]

In 1817, in evidence to a Parliamentary Select Committee, Taylor still described his 'principal pursuit' as that of 'manufacturing chemist at Stratford in Essex'.[10] This was not to last much longer however. In September of that year he further revived his old Tavistock connections by becoming a member of the seven man trustee committee of the canal company, following on John Martineau's retirement from that post, and within a month was again in temporary residence in the area, instigating work to overhaul the port facilities at Morwellham and encouraging greater output at the canal's mines to take advantage of a sudden increase in copper prices.[11] By 1820 he had acquired land in Tavistock and was engaged in building offices and other premises for mining business around a square which still bears his name and in 1821 he promoted a new company to re-open his old mine, Wheal Crowndale, on highly advantageous terms which he had re-negotiated with the Duke of Bedford.[12] With the formation of this company, he completed his reinvolvement with all of his old Tavistock interests, although now as capitalist and entrepreneur and not simply as paid manager. Perhaps this had been his intention all along.

IV Ecton Hill and Grassington Moor

Taylor obtained his second appointment as a mineral agent in 1818, this time to the powerful Duke of Devonshire, taking charge of his mineral property around Ecton Hill in Staffordshire and on Grassington Moor in Yorkshire.[13] As at Lord Grosvenor's Halkyn mines, his first action was to take steps to replace the existing managers with Cornish captains of his own choice. The ancient Ecton Copper Mine, which the Duke was working on his own account, was then being managed by George Cantrell who Taylor found to be much 'in want of science' and he was soon supplanted by Captain Goldsworthy from Cornwall.[14] Under their combined direction they introduced all that was best and most modern in Cornish organisation and mining and dressing technology. They set in hand the large-scale reworking of all tips and waste ores, which were previously regarded as worthless, using Cornish water-powered stamps and the latest tramway handling systems, introduced into the area for the first time. Although continuing with the local calcining of the wastes, they closed down the neighbouring Whiston smelting works and transfered the reduction of all new copper concentrates to Swansea, which was the smelting and manufacturing centre for all Cornish ores. They also installed underground a large new overshot waterwheel to replace the old, less efficient water-pumping engine. Perhaps the most important development, however, was the substitution of the Cornish *tribute* system of employment for existing *cope* contracts.

Both types of contract were essentially a loose form of piece work, but in the tribute bargain the miners' earnings were tied more closely to the quality as well as the quantity of their work. As Taylor was later to explain in a lecture to the Geological Society, the proper and efficient organisation of a mine's labour force was the most important single factor in ensuring its economic operation: more important than any single advance or improvement in mining technology. Thus he declared that:

> Important as the improvements are, which we have contemplated in the instruments which the progress of physical science has placed in our hands, those which relate to the government of large bodies of workmen, to the inducement to active enterprise on the part of the labouring miners, to the removal of difficulties in their way, or of placing them in circumstances most favourable to effective exertion, are even more important, and to this may be added the judicious application of those very inventions which have been noticed. It must be recollected that, after all, the great expenditure in mining is for manual labour, and that we have no means as yet devised for penetrating the rocks which contain mineral treasures but those afforded by the patient and unremitting labour of a great number of men. The regulation, therefore, of this force, and its due application is, after all, more important to the success of mines than even the most ingenious mechanical expedients. As an army would undoubtedly fail, however well provided with the most perfect artillery, and all the best constructed implements of war, unless the men of which it might be composed were well directed, their efforts well combined, and their courage assured by reasonable prospects of success, so in mining we may collect and apply the most complete mechanical arrangements, but if the greater power of manual labour be not wisely directed, no beneficial results can be expected.

It was part of Taylor's peculiar mining genius that he realised and implemented these policies more completely than most of his contemporaries. During these same years he also introduced 'the Cornish system' into the mines of which he had charge in Flintshire and Yorkshire and by the mid-1830s had extended it into Cumberland, Cardiganshire and Ireland, where it underpinned most of the reconstitution and development planning. Although he admitted that it was not done without some initial opposition, such as the troubles at Halkyn in 1822, he claimed that, 'the men soon found that fair play was given to all, that it left no room for undue preferences and that by it the more skilful and industrious were . . . better rewarded'. He concluded that, 'I know of no case where the men would wish to return to the old plan'.[15] The elegance of the system and the way in which it directly related earnings to effort and skill, also strongly recommended it to the political economists of the day. Thus Charles Babbage, a close friend and an admirer of Taylor's business and technical expertise, quoted the principle of the system as a model for all employers of labour in his monumental work *On the Economy of Machinery and Manufactures* (1835).

On Grassington Moor, John Taylor found that the operations at the Duke of Devonshire's mines had been suspended for some years while waiting on the arrival of a deep drainage and drawing adit, known as the Duke's Level. He immediately set about restoring production by reducing the size of the level, to speed up driving, and installing heavy pumping equipment, powered by a 50ft diameter overshot waterwheel, which would cope with the drainage problems until the adit arrived. To take charge of the day-to-day operations of the mines he again introduced his own Cornish captains; John Barratt, later joined by Stephen Eddy. Like his father before him, Barratt had been manager at Wheal Friendship and the success which he and John had there in exploiting water power for underground and surface operations inspired a new complex system of installations on the Moor. When finally completed this system consisted of nine dams for water storage and regulation, feeding through about $6\frac{1}{2}$ miles of leats to at least eight large waterwheels providing power for pumping, winding and dressing operations.[16] To make maximum use of these improved power supplies, pumping and winding arrangements were reviewed and large concentrated dressing floors were set up. These made maximum use of imported Cornish stamping and crushing machinery together with improved mechanical jigs, buddles and railway lines, the latter to minimise handling costs and ensure a smooth through-put of ore. Giving Barratt the greater part of the credit for this new integrated system, Taylor later praised it as 'the most systematic plan of ore dressing that I know'.[17] Taken together, these improvements produced a large increase in the output and profitability of the Duke's property.

In the mid-1830s Taylor transferred Captain Barratt from Grassington to his newly-acquired Coniston copper mines in Lancashire and here again their combined abilities produced a story of restored production and prosperity. By the end of that decade Barratt was becoming wealthy in his own right and bought a share in the ownership of those mines. Still later, in 1855, having finally left the service of the ageing John Taylor, Barratt played a major part in the re-opening of the great Hodbarrow iron mine in Furness, which was to become the largest single producer in Europe.[18]

V Cornwall

Not all of Taylor's territorial expansion during these years was achieved by acquiring posts as mining agent. In 1818, the same year that he took up his position with the Duke of Devonshire, he also launched a company to work Wheal Virgin, Wheal Fortune, and other large mines near Gwennap in Cornwall, under the name Consolidated Mines. This was not his first excursion into that county, having already been involved at Wheal Pembroke, in the St Austell district, but it was his largest enterprise so far. Indeed, it was one of the most ambitious projects of its day.[19] It had already defeated the great organisational talents of well-established entrepreneurs, such as the Williams of Scorrier, and its fabulous future success was to make it one of the most strategic turning points in Taylor's career. It confirmed him as the country's leading exponent of mining affairs and endowed him with an international reputation. The original company was formed with 100 shares and a subscribed capital of £15,000. Taylor introduced most of the other investors, mainly from the London area and through his close control of every aspect of management gave the concern the appearance almost of a 'private adventure'.[20]

When the company took over the mines in 1818 they had all been idle for some years. By mid-1819 they were restored to production and by 1822 the group had an output of more than £80,000 worth of ore per annum, establishing them as the largest producer in Cornwall, a position which they retained for the next twenty years.[21] The sustained high level of output meant that by late 1837 the company had sold over a quarter of a million tons of ore valued at nearly £2 million. During those 18 years they had incurred working costs of almost £1½ million, about two thirds of which was wages for

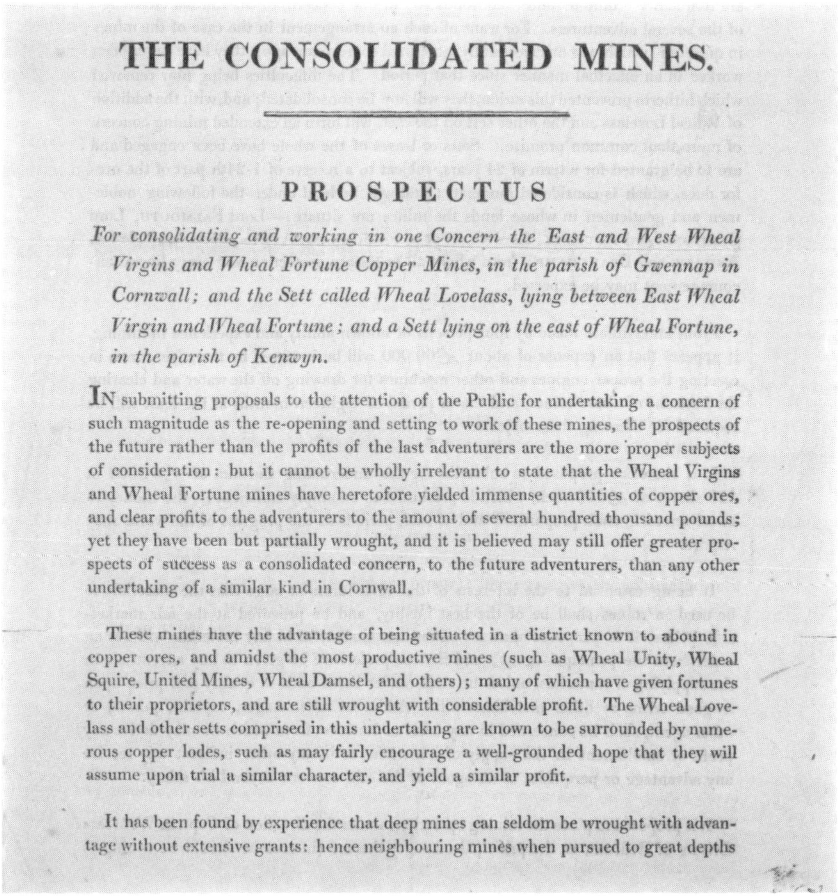

THE CONSOLIDATED MINES.

PROSPECTUS

For consolidating and working in one Concern the East and West Wheal Virgins and Wheal Fortune Copper Mines, in the parish of Gwennap in Cornwall; and the Sett called Wheal Lovelass, lying between East Wheal Virgin and Wheal Fortune; and a Sett lying on the east of Wheal Fortune, in the parish of Kenwyn.

IN submitting proposals to the attention of the Public for undertaking a concern of such magnitude as the re-opening and setting to work of these mines, the prospects of the future rather than the profits of the last adventurers are the more proper subjects of consideration: but it cannot be wholly irrelevant to state that the Wheal Virgins and Wheal Fortune mines have heretofore yielded immense quantities of copper ores, and clear profits to the adventurers to the amount of several hundred thousand pounds; yet they have been but partially wrought, and it is believed may still offer greater prospects of success as a consolidated concern, to the future adventurers, than any other undertaking of a similar kind in Cornwall.

These mines have the advantage of being situated in a district known to abound in copper ores, and amidst the most productive mines (such as Wheal Unity, Wheal Squire, United Mines, Wheal Damsel, and others); many of which have given fortunes to their proprietors, and are still wrought with considerable profit. The Wheal Lovelass and other setts comprised in this undertaking are known to be surrounded by numerous copper lodes, such as may fairly encourage a well-grounded hope that they will assume upon trial a similar character, and yield a similar profit.

It has been found by experience that deep mines can seldom be wrought with advantage without extensive grants: hence neighbouring mines when pursued to great depths

11 Front page of Taylor's prospectus for the Consolidated Mines, Gwennap, Cornwall. *(J. A. Robey).*

their labour force of more than three thousand men, women and children, and they had driven over 63 miles of shafts, levels, winzes and rises. After paying off their initial capital and interest charges in 1829, they had distributed nearly a quarter of a million pounds in clear profits to their shareholders.[22] It was clearly an enterprise which compared in scale and turnover with the largest in any sector of industry and commerce and, together with other mines in the Gwennap area acquired by Taylor and his associates in the early 1820s, it completely dominated the district.

Yet again the success of the company was founded on heavy investment in the most up-to-date mining and dressing technology combined with astute financial management. Everywhere John looked for opportunities for mechanisation and economy. From the very re-opening of the mines, regular searching experiments were made with different types of steam pumping engines to determine the most efficient designs and methods of operation. It was soon found that compound engines, of the type designed by Arthur Woolf and strongly favoured by most mines during the first two decades of the nineteenth century, were high in power but were less efficient in the consumption of fuel than the new generation of large single-cylinder engines. Since fuel costs were a large and increasing item in the overall expenses of the deepening mines, John decided clearly in favour of installing single-cylinder engines, setting a precedent which was to be followed by mines and other major pumping establishments throughout Britain and abroad.

Notwithstanding this apparent conflict of interests, John Taylor and Arthur Woolf became close friends and Woolf took up the position of chief engineer at Consolidated, from which he retired with a lifelong pension in 1830, being replaced by the brilliant young engineers John Hocking and Michael Loam.[23] By 1836 the Consolidated Mines had installed eight large pumping engines and a similar number of steam drawing-engines. These were complemented by a 48ft diameter pumping waterwheel, a 40ft diameter machinery wheel, five smaller wheels for pumping and crushing and numerous horse whims above and below ground.[24]

At first John implemented his new technology and investment programmes at Consolidated himself but later he collaborated with his second son, Richard, who he installed in the senior managerial position. The introduction of dressing machinery to Consolidated's dressing floors provides a good illustration of this managerial system. In 1806, when at Wheal Crowndale, John had devised a mechanical crushing mill for reducing mixed veinstuff to smaller separable pieces. With this device he had hoped to overcome temporary labour shortages in this initial stage of the dressing process and to increase output to take advantage of the current high prices for copper ore. Fortunes soon changed, however, and it was not until after 1823, at Lord Grosvenor's Halkyn mines in Flintshire, that he was able fully to develop and perfect this machinery. By 1831 he was so convinced of its usefulness that he called on Richard to install it at Consolidated. Here it proved such a resounding success that more than half a century later Robert Hunt could still write that, 'From that time to the present, the roller crushing mill has formed the principal apparatus for reducing mixed lead and copper ores.'[25]

Having broken a bottleneck in the initial reduction of the ore, John and Richard immediately took steps to mechanise the equally labour-intensive separation and classifying processes. After seeing how one particular method of separation (jigging) had leant itself to mechanisation in the North of England lead districts, John had for some years been interested in introducing and improving on this technique in the South-West and had encouraged experiments with jigging machines at his Tavistock mines during the late 1820s. At the same time, at Fowey Consols, Consolidated's great rival, Thomas Petherick had been working on an entirely new type of jigging machinery and was beginning to contemplate its full-scale introduction. Accordingly in 1831, the same year as the introduction of the crushing mill, Richard also began the construction of mechanical jiggers. There were some initial problems, particularly from the labour force who feared for their jobs, but these were quickly resolved and the machinery was soon an established success.[26] From these two examples at least it would appear that John Taylor's reputation for technical expertise rested not simply on the introduction of advanced Cornish mining technology into more backward areas, but on some considerable inventive talent of his own and an open-minded cross fertilisation of ideas: taking the best techniques and methods of every district and applying and improving on them whenever they appeared suitable.

As well as increasing the efficiency of mining and dressing operations, Taylor also looked for improvements in the facilities for transporting ore and supplies between the mines and the coast. The mines' traditional connections with the outside world had been through the ports of the north coast, but Taylor looked to a closer and more direct route to the south, using the river Fal. In his attempts to reduce costs, John was probably hoping to emulate his earlier success with the Tavistock Canal, but on this occasion he chose not an all-water route but the cheaper, more flexible and currently popular, tramway system. Interest had been shown in such a project even before the mines were restarted and in 1823 formal Parliamentary permission was sought for a rail link from the Gwennap mines and others around Lanner and Redruth to Point Quay on the Fal, to be known as the Redruth and Chacewater Railway. An Act of Parliament was received on 17 June 1824 and a separate railway company, formed from Taylor's London associates, was launched with a capital of £22,500. By the end of 1825 more than nine miles of line was in operation, the main traffic being the carriage of ore out and of coal in. Although the original horse-drawn trucks were not replaced by steam locomotives until 1854, the line succeeded in considerably reducing costs and over the first thirty years of its operation produced a profit for the company of more than £70,000.[27]

Notwithstanding the excellence of the mechanical and transport innovations which Taylor introduced, his greatest improvements at Consolidated were managerial, not technological. He made it an express condition of the constitution of the company that contracts for the supply of all materials to the mine should be awarded in fair and open competition. This was a unique provision in an industry where such contracts were customarily awarded as an important perquisite to major shareholders and mine officers. It represented a serious attempt to place mine finance on a regular footing and to avoid the exploitation which had transformed many potentially profitable workings into defunct debtors. To that large body of mineral owners, investors and mine captains who took interests in mines not for the dividends which they produced, but for the advancement of their own private trading functions, it also represented a major threat to the status quo and an example which, if allowed to succeed and be followed by others, would mean serious financial losses. Accordingly they conducted a running battle with Taylor and the Consolidated company throughout the 1820s and early 1830s, finally bringing the issue to a head when the company applied for a renewal of its lease in 1836. In the negotiations for renewal, which dragged on for the next three years, the mineral owners insisted on a one-third interest in the company and the appointment of a similar proportion of their nominees to managerial posts. Taylor saw this as a clear challenge to his managerial policy and sold up the company rather than agree to the new terms.

Messrs Williams and their friends, Taylor's old protagonists, purchased the company's 100 shares for £90,000 and took over operations in July 1840. They continued working at the mines for another 18 years but annual production and profits gradually fell off; the value of production over these years being just over a quarter of that during a similar period under the old company.[28] Recalling these events on Taylor's death in 1863, a correspondent of the *Mining Journal* wrote,

> The leases . . . expiring, these mines were ruthlessly torn from his hands by parties who, not content to receive their portion of the regular 'golden eggs', determined to seize the 'goose' and have the whole: when lo! on opening the 'goose' the 'golden eggs' had vanished.[29]

It is notable that John Taylor was far from defeated by this reversal at

Consolidated and the principle of putting supplies contracts to tender was adhered to, with continuing attacks, at all of his mines in all districts until long after his death. John also retained control of the Redruth and Chacewater Railway Company for the rest of his working life; at first directly and later, after 1850, as General Manager, with his son Richard again taking charge locally as 'Treasurer and manager in Cornwall'.[30]

VI Alston Moor

In April 1823, when already involved in the supervision and management of mines in Devon, Cornwall, North Wales, Staffordshire, and Yorkshire, John Taylor sought still further geographical expansion by writing to the Board of the Commissioners of the Greenwich Hospital offering to survey and report on the best development of their rich lead-bearing mineral property on Alston Moor.[31] He was commissioned for that purpose and later that same year travelled to Alston Moor in company with Robert McAdam, who was engaged to survey the roads of the district. He conducted an exhaustive twelve-day investigation and submitted a comprehensive detailed report.

The report confirmed the Commissioners' existing practice of leasing their mines to companies of independent adventurers, rather than directly working them themselves and agreed with the general conditions of the leases. Taylor also favoured the existing practice of letting the mines at a standard rate of duty rather than at rates varying with the value of the lodes worked, but added that these general duties were currently higher than in most other districts and it might be necessary to vary them should the mines decline in productivity, incur increasing costs, or see a reduction in the value of their ore. The only major new suggestions which he made concerning the arrangements for leasing the mines were that the lessees should be required to keep their underground workings as clear as possible and arched with stone rather than less-durable timber, so that others might have easy access to the workings if they should be abandoned; that careful plans of all workings should be kept;[32] and that the numerous leases to the London Lead Company should be rationalised.

Taylor found that the working of the mines, both above and below ground, was generally satisfactory, particularly at the extensive London Lead Company mines but suggested that improvements were possible everywhere. He declared,

> I am of opinion that more may be done with good effect, and particularly in some of the Mines, by a better system in laying out the portions of the vein, for working in a more regular way, by revising the modes of agreeing with the workmen, and applying constantly the principle of making their advantage to be concurrent with the discovery, and the most effectual exhaustion of the Ore, by arrangements conducive to the convenience and health of the persons employed. By more care and skill in the washing and preparation of the Ores for Smelting, and particularly in regard to such parts of their processes which incur waste, and which I think are too generally to be found. All these things are however more immediately the affair of the Mining Tenants, and though they are highly important to the interests of the Estate, yet I am well pursuaded, that any attempt to enforce regulations would be attended with great chance of failure, and when abortive would do hurt instead of good.

Taylor added, therefore, that the improvements should not be incorporated in a revised form of lease but that the mine adventurers should simply be encouraged towards better practice. To this end he suggested that the Hospital should appoint a highly qualified Inspector of Workings, to advise on how the mines might be improved and to report to the Hospital officials on the progress made.

The main flaw which Taylor detected in the Hospital's operations related to the ore which it received as royalty from the mine adventurers and which it occasionally purchased from the smaller mines to save them from the inconvenience of marketing directly. To improve the saleability of this ore and supposedly to increase their overall profit, the Hospital currently smelted it in its own works at Langley. Taylor observed:

> As I have expressed my opinion, that it would be unwise for the Hospital to engage in working their mines, so I should doubt how far it is advisable that they should commit themselves to the detail of a manufactury like Smelting, which requires a perculiar kind of knowledge and the attention to many minute particulars, so that in general, individuals have better succeeded in it than large companies . . . if the Hospital could at once sell all their duty ore, and the smaller Companies of their Miners could get a fair value for what they produce, I should have no hesitation in saying that such sales would be better than the complicated Machinery of Smelting Houses etc.

There were, however, major problems confronting the implementation of these suggestions. As he continued:

> At present, I see no way in which either of these could be done. In some other lead districts, smelting the Ore is a distinct business from Mining, and is carried on by another set of people, and in accordance with the proper division of labour, I believe this is the best practice. In some of these places the smelters are sufficiently numerous and respectable, to ensure both fair competition to regulate the Prices of Ore, and to guarantee payment to the sellers, and if this

> were the case in the North, there would be no difficulty in changing your system, if you thought it advisable. Such a body of purchasers of Lead Ore, are, however not to be found within reach of the Aldstone Moor Mines, the practice of the country has always been for great proprietors, or their principle lessees, to smelt their own Ores, and no opening therefore existed for the establishment of Smelting Works to depend on purchased Ores . . . The Officers of the Hospital therefore seem to be compelled to convert the Ore into Lead, as the only saleable form, and at a period of depression in the market they probably preserved the poorer miner from ruin by undertaking to purchase his ore also . . . As the thing now stands I do not see how you can alter the present mode, as any difficulty thrown in the way of the poorer Miners in disposing of their Ore, would depress their exertion, and injure the Mines, and the duty Ore must evidently be made saleable in the best way it can.

In the immediate future Taylor thought the best course to be the improvement of the smelting operations by developing a better system of price fixing; observing the outcome of experiments currently being conducted by the London Lead Company in Teesdale on the possible advantages of changing from the existing blast hearths to reverberatory furnaces; appointing an Assay Master to ensure that the maximum amount of metal was being extracted from the ore; and perhaps exploring the possibilities of renewing and re-opening the zinc smelting department of the Langley works.

The final part of the general report was devoted to considering the 'large Expenditure' on that 'unfortunate speculation', the Nent Force deep drainage and exploratory level, which first started driving in 1776. Notwithstanding the great losses which had been accumulated over the years, Taylor was not too critical of this project since he appreciated that,

> there were fair reasons to expect a different result, the intersection of so large a tract of Mineral Country, and the exploring deep beds of Lime Stone similar in many respects to that which has produced so large a proportion of the Lead raised at Aldstone Moor, appear to me to have warranted the undertaking, and to justify those who so long ago recommended it.

On a similar principle he drew short of suggesting the immediate termination of the venture, advising instead continued exploration in different directions and at a higher level,

> first, because the portion of ground which will thus be explored appears to me to be situated where discoveries are most to be expected; and secondly, because in the event either of a continuance of unfavourable appearances, or of an alteration for the better, time is very important, either by coming to a resolution to stop the expenditure, or arriving at a beneficial result quickly.

Taylor suggested that part of the expense of this continued activity might be recovered from the adventurers through whose leases the level passed, but reminded the Commissioners that the mines were already paying very high royalties and that they should not be too demanding in case they began to discourage activity.

In conclusion, Taylor paid tribute to the general 'discretion and sound judgement' of the Hospital receivers and managers and attributed to them 'in a great degree, the present prosperity of the Mines, and the spirit with which the Adventurers carry them on'. His only major suggestion was that their number should be complemented by the appointment of an Assay Master to keep check on the output and effectiveness of the smelting operations, and an assistant to the Moormaster, preferably recruited in Cornwall or some other area of advanced mining techniques, to advise and encourage the adventurers on the best conduct of their mines. As for the future of the Estate, Taylor was optimistic.

Standing back from the immediate short-term problems and showing a depth of perception based on a long and disparate association with mining affairs, he wrote:

> Large estates of this kind . . . have commonly their periods of comparative richness and poverty, and these alternate from a variety of circumstances. Old Mines seem to be exhausted and are abandoned, a desponding feeling is produced and the spirit of enterprise is depressed, and a consequent reduction takes place in the value of the produce; this may go on and has done so in many instances for a considerable length of time, but it has generally happened that either an improved method of research, accidental discoveries, or the application of more effectual or more economical methods of working have renewed the spirits of Adventurers, and as it were given new life to the undertakings necessary for opening productive mines, from these a restitution, or very often (and I might say more generally) a great increase of former Revenue has arisen. Few mining districts in England, perhaps hardly any important ones, can be said to have been *exhausted*, even though the history of some and a continued series of working may be traced to remote ages. The Mines of Aldstone Moor, have fluctuated in the same manner and may be expected to do so again, but I have no doubt of their continuing a very long period to produce on an average of years a great Revenue. The older concerns are not rich at present, and their prospects in depth are limited by the peculiarity of the stratification, unless the great level should discover bearing veins in the deeper beds, but if they wear out and decline, other veins still unknown will probably be found, and keep up or restore the productiveness of the Estate; such an instance is to be found in the case of Hudgill Burn, and indeed it is one of the common uncertainties of mining, that while disappointment often occurs in the pursuit of a favourite object, unexpected discoveries are made which compensate the losses.[33]

Taylor's recommendations for improving the administration of the estate and the working of the mines were substantially accepted and henceforth he was engaged as *Inspector of the Hospital's Mines,* receiving a retainer to conduct annual surveys of the estate and report directly to the Commissioners. On his recommendation Thomas Dickinson, the son of the Hospital's chief administrative officer, was sent to Cornwall to study the latest mining practice so that he might assist and later succeed his father on his retirement. Taylor's nominee, H. L. Pattinson, who was soon to earn an international reputation for a brilliant cost-saving innovation for refining silver, was also appointed as the first full-time Assay Master. During the late 1820s and early 1830s John negotiated limited royalty reductions for some mines to ensure proper investment and long-term development, acted as arbitrator in disputes between mining companies, and partly redrew the Hospital's mining regulations and leases. Further experience and changing conditions all persuaded him to revise his earlier advice and convince the Commissioners to rationalise their activities by selling off the smelting mill at Langley and agreeing with the lessees to receive their royalties in money value rather than ore. Time also solved Taylor's early quandary about the wisdom of continued investment in the Nent Force Level. Finally completed in 1839, it had some effect in stimulating other trial and drainage projects but it never produced any major new discoveries and contributed little to the alleviation of drainage problems, most of the larger mines having long-since gone below its level. As Taylor had suspected, it probably never repaid the large sums of money invested in it.

During the 1840s and 1850s Taylor's work for the Hospital gradually dwindled and was probably increasingly conducted for him by his eldest son. Finally, in 1859, at the age of 80 and probably not having visited the district for many years, he officially retired from the position.

REFERENCES

1 *Dictionary of National Biography.*

2 The gas which they derived from oil was claimed to be considerably more pure than that from coal and had it not been for improvements in methods of purifying coal gas, it might have become of considerable importance. See A. & N. L. Clow, *The Chemical Revolution* (1952), p436. For a detailed description of their process, see *Quarterly Journal* (1820), p120.

3 BM Add Ms 34612, Letter 8 Oct 1819.

4 B. Woodcroft, *Alphabetical Index of Patentees of Invention* (1854), p559.

5 Report of the Select Committee on Steam Boats, *British Parliamentary Papers* (1817) **VI,** Evidence of John Taylor, pp14-17 (236-9).

6 See W. J. Lewis, *Lead Mining in Wales* (Cardiff, 1967), pp208-10 and A. H. Dodd, *The Industrial Revolution in North Wales* (Cardiff, 1933), p402.
7 John received further gifts of 100 gns from the company in 1821, '22, '23, '24 and '27.
8 In 1826, for example, he instigated the sinking of an additional inclined haulage shaft at the mine. See F. Booker, *Tamar Valley*, p131.
9 *Wheal Friendship Minute Books* in the DRO on loan from Mr Justin Brooke.
10 *BPP* (1817) **VI** (Evidence), p14.
11 DRO Bedford Collection, Tavistock Canal Papers No 21 and L 1258 No 83.
12 Royalties for this new venture were to be only 1/24th of output until £1,000 clear profit had been divided and then increased to 1/15th, which was still lower than the 1/12th average in the area. See DRO L 1258 No 83, Letter 7 Feb 1821.
13 *Mining Journal*, 22 Jan 1838, p27.
14 J. A. Robey and L. Porter, *The Copper and Lead Mines of Ecton Hill, Staffordshire* (Moorland Publishing Co, 1972), pp41-2.
15 J. Taylor, 'On the Economy of Mining' *Quarterly Mining Review* **X** (1837), pp261-2 and reprinted in R. Burt (ed) *Cornish Mining* (Newton Abbot, 1969), pp34, 35, 39.
16 See A. Raistrick, *Lead Mining in the Mid-Pennines* (Truro, 1973), pp114-5 and *Trans Newcomen Soc* (1955) **29,** pp179-93 for a detailed discussion of these installations.
17 R. Hunt, *British Mining* (1887), p697.
18 A. Harris, *Cumberland Iron* (Truro, 1970), p17.
19 D. B. Barton, *A History of Copper Mining in Cornwall and Devon* (2nd Ed, Truro, 1968), p58.
20 *MJ,* 16 Feb 1839, p52.
21 C. C. James, *A History of the Parish of Gwennap in Cornwall* (Penzance, 1949), p205.
22 *MJ,* 16 Feb 1839, p52 and Sir Charles Lemon, 'The Statistics of the Copper Mines of Cornwall' *Journal of the Statistical Society of London* **I** (1838), pp65-84, reprinted in R. Burt (ed) *Cornish Mining,* pp49-82.
23 D. B. Barton, *The Cornish Beam Engine* (Truro, 1966), pp40-1, 142-3 and T. R. Harris, *Arthur Woolf: The Cornish Engineer 1766-1837* (Truro, 1966), p93.
24 Sir H. de la Beche, *Report on the Geology of Cornwall, Devon and West Somerset* (1839), p600.
25 R. Hunt, *British Mining* (1887), p694.
26 *Ibid.*
27 D. B. Barton, *The Redruth and Chacewater Railway 1824-1915* (Truro, 1966).
28 C. C. James, *Gwennap,* pp205-6.
29 *MJ,* 19 Sept 1863, p662.
30 D. B. Barton, *Redruth and Chacewater Railway,* p23.
31 These Alston Moor estates had been confiscated by the crown from the Earl of Derwentwater for his support of the 1715 rebellion and were assigned to the Greenwich Hospital as a source of income in 1735.
32 The need for carefully kept records and plans of underground workings was one which Taylor publicly advocated and struggled for throughout his active life.
33 *Report of a Visitation of the Greenwich Hospital Estates in 1822 and 1823,* BM 796 1 30.

CHAPTER 3

The Mexican Ventures

In 1824, with several consultancies and nearly forty large mining companies already under his direct control, John branched out yet again. This time, however, it was not into another domestic mining district but abroad, into the 'rediscovered' mines of Mexico.[1] Following the revolution to overthrow Spanish colonial rule, the Mexican mineral owners looked abroad for capital to re-open the once-rich mines on their property, which had fallen into disuse during the revolutionary struggle. The Count of Regla, the most powerful of this group and the owner of the once fabulously rich Real del Monte silver mines, commissioned the English firm Rupert Staples and Company in Mexico as his agents in this business. In 1823 they transferred it to their London partner, Thomas Kinder, who having heard of the 'great success and experience' of John Taylor in mining ventures, approached him with the proposal.

Taylor, as topical as ever, was already well acquainted with the history and potential of the Mexican mines. As early as 1819 he had acted as consultant to a British firm that was contemplating sending a steam engine for use in a Mexican mine and in 1824 he was to publish, with notes and informed introduction, selections of the work of his friend Baron von Humboldt, then Europe's leading expert on the geography and economy of South America.[2]

With enthusiasm on both sides, Kinder and Taylor soon agreed that a company should be formed to work the Regla property and from their association the British Real del Monte Company was born. It took the form of the currently fashionable, but as yet illegal, joint stock company, with an original capital of £200,000 divided into 500 shares of £400 each.[3] Taylor reserved the right to distribute 300 of these shares between himself and his group of London associates and Kinder claimed 200 shares, 50 of which he might sell to interested parties in Mexico before being required to pay the first instalment on their purchase price. After two meetings of the shareholders it was decided to empower Taylor to conclude leases not only with Kinder for the Regla property but also with Colonel John Murphy, acting on behalf of his brother Thomas, for mines which he owned in the neighbouring Moran district. It was also agreed to make an immediate call of 5 per cent of the purchase price of the shares and to instruct the company's solicitors – Martineau and Malton, part of the extended Taylor-Martineau family enterprises – to prepare a deed of settlement or promote an Act of Parliament for the proper incorporation of the company. On 6 March 1824 Taylor signed 20 and 21 year leases for the two properties which, among other things, committed the company to a programme of investment in the latest steam operated machinery at the mines.

Less than three weeks later the first advance party of fifteen officers and technicians sailed from Liverpool for the Real, led by James Vetch, who was to serve as the company's chief commissioner in Mexico, and two other commissioners, John Rule and Vincent Ravafinoli, an Italian friend of the Count of Regla. On 16 August of the same year the company was given legal corporate status by the adoption of a deed of settlement. This also established its structure of government and rules of procedure which, although later hailed by the Lord Chancellor as a model of its kind, was to cause the company continual managerial problems.

Some time before the advance party left for the Real, Taylor and some of the other investors in the company also became interested in the prospect of arranging leases of the mines in the adjacent Bolanos district, which had been described by Humboldt as the fifth richest in Mexico. Rather than incorporate these leases in the existing company, Taylor and his immediate group of financial backers sought to obtain these leases for themselves and to establish an entirely separate Bolanos Mining Company. Accordingly, Taylor instructed Captain Vetch to make appropriate enquiries on his arrival in Mexico. Unfortunately this produced an almost immediate split within the Real del Monte Company as those shareholders outside the Taylor group, led by Thomas Kinder, thought themselves defrauded. The dispute quickly led to a lawsuit in which Kinder claimed that he had been denied a share in the Bolanos enterprise and asked that the Court of Chancery direct Taylor and his friends to grant him a fair portion of the Bolanos mines and to dissolve the Real del Monte Company if that was necessary.[4] Fortunately, the Lord Chancellor accepted Taylor's case almost completely and in his judgment endeavoured to produce an even wider separation of the companies than achieved so far; attempting, for example, to prevent the joint employment of staff in Mexico.

Whatever the resolved legal situation, the stories of the Real del Monte and Bolanos companies continued to be closely related throughout their existence. It was not until the end of the 1820s that men such as Vetch and Charles Tindal relinquished the joint management of the two enterprises at the Mexican end of the operations and as late as the mid-1840s one-fifth of the Real del Monte shareholders also held shares in Bolanos and one-fourth of the Bolanos shareholders in the Real del Monte Company. Throughout its existence the Bolanos enterprise was smaller than the Real but generally proved more profitable, realising more than £1 million. From the early 1840s both companies followed a similar downhill path and by the end of that decade both had gone into liquidation.

In the Real del Monte and Bolanos enterprises Taylor hoped to repeat on an international scale what he had already achieved at home: to use his unparalleled financial and managerial expertise to revive the once profitable mines of a now moribund district by a carefully planned injection of capital and the most up-to-date mining and dressing technology. It was a formula that had already produced almost immediate and unqualified success in the copper and tin mines of the South-West and the lead mines of North Wales and the Pennines and there was no reason in principal to suppose that it would fail in the silver mines of Mexico. In practice, however, it was soon found that although the problems of international activity were in no fundamental way different from those of moving into a new domestic district, they were of much greater magnitude and required different and more flexible solutions. Taylor's failure fully to appreciate this and to adapt his policies accordingly – together with a large amount of 'bad luck' which had always figured prominently in the success or failure of all mining operations[5] – was to prove the undoing of both his Mexican ventures.

This is nowhere better illustrated than in the administrative arrangements of the Real del Monte company and their shortcomings. Conforming with his now usual practice of retaining direct overall supervisory control of the mines with which he was connected, Taylor took steps to ensure that under the terms of the deed of settlement setting up the company, complete authority for the management of all workings should be vested in London, in the hands of a Court of Directors. The officials in Mexico, rather like the captains at his other mines in England and Wales, would be neither managers nor directors, but simply commissioners acting almost entirely on instructions from London with little delegated authority of their own. Even this was not enough for Taylor, however, and he quickly persuaded the Court of Directors to establish for him the special post of Manager with which they would share their authority equally. Taylor thus had almost complete control over the running of the mines, receiving reports and communications from Mexico and submitting reports to shareholders independently of the Court of Directors.

Although this procedure may have worked well within the mining districts of England and Wales, safeguarding the various mines from the perils of incompetent managers, it broke down under the greater distances and communications problems with Mexico. Firstly, in England and Wales Taylor knew his mines well, visited them frequently, and was intimately acquainted with the character of their underground development and the methods of dressing. By contrast, in more than twenty years of the Mexican enterprise he never found the time to visit those mines and as a result, perhaps never fully understood the special problems which they presented. Secondly, in England in the early nineteenth century, even before the railways and the telegraph revolutionised communications, letters, information, instructions could be conveyed from one end of the country to the other within a matter of days. Thus even when away from his mines, Taylor could receive intelligence and return instructions quickly. For the Mexican mines, however, the time intervals needed to be reckoned in months not days. Even under good conditions, letters written in London or Mexico required between six weeks and two months to reach their destination and bad weather, blockade or other difficulties could produce a delay of up to several weeks more. Even if a decision was taken in London immediately on receipt of a communication, which was itself unusual, a similar time period would elapse before it would be acted upon at the mine. It would not be unusual, therefore, for a time period of six months or more to intervene between the commissioners in Mexico seeking advice on a problem and the receipt of instructions from England, during which time conditions might have again changed dramatically. Before long this situation produced basic frustrations and antagonisms between Mexico and the London headquarters which served only further to aggravate the situation.

In the same way that distance created administrative difficulties, it also caused serious supply problems. One of the corner stones of Taylor's success in British mining was economy in the purchase of mine supplies, largely from local dealers. In the Real, with no established industrial infra-structure, this was virtually impossible. Almost everything, from pen nibs to steam engines, had to be brought from England. This entailed not only a long and arduous sea crossing – with the constant prospect during the early years of the only convenient ports of disembarkation being blockaded by the Spanish – but also expensive overland haulage along more than 200 miles of poor roads from the coast to the high sierras. In the lower coastal areas sickness took an immediate toll of the newly arrived Europeans and along much of the route the supply parties were subject to constant attack by bandits, particularly if it was known that they carried specie for paying the miners.

The Mexican authorities, who might have been expected to help and encourage the development of the mines, contributed little and their actions often served to frustrate the companies' own efforts. They imposed awkward regulations on where the company might land its machinery, left the cost of road improvement entirely to the company, made no effective effort to solve the bandit problem, and on several occasions became involved in wars which effectively disrupted the companies' sea communications. Regulations controlling the minting and export of the companies' silver output also carried constant difficulties and the government's claim to a monopoly of the manufacture and supply of gunpowder, which they could seldom fulfil, nearly brought the mines to complete closure on several occasions and obliged them to rely on supplementary supplies, sometimes purchased illegally.

Another cornerstone of Taylor's domestic success was the innovations which he introduced in the employment of the labour force. These also proved extremely difficult and costly to repeat in Mexico and were never fully achieved. Labour was scarce in the remote mining districts and throughout the years of activity enjoyed a strong position vis-à-vis the companies. They were accordingly less susceptible than their British counterparts to pressures to accept unpopular conditions or changes in terms of employment. The Mexican miners, or 'barrateros', also had a strong group consciousness and a history of militancy and violence in defence of their traditional 'partido' contract, a type of profit-sharing scheme which enabled them to take a large part of their earnings in the form of dressed ore. With Taylor now used to brushing aside all opposition in having his own way, all the ingredients were present for a series of bitter conflicts at the mines. Between 1827 and 1845 there were six violent and financially crippling clashes with the barrateros at the Real del Monte mines, all the result of attempts to reduce what the company regarded as unduly high wages by the modification of the partido to the tribute system. The company lost all of them, though by the end of the day the partido system had been largely eradicated mainly because of declining interest by the barrateros as the smelters turned away from the *type* of ore which it produced for them. It was replaced, however, not by the tribute bargain but by tut, in which the miners shared fewer of the risks of unfavourable changes in the nature of the ground which they were working. To the company's displeasure, overall earnings continued to be high – but perhaps no higher than elsewhere in the district – and in the final declining years of the mines they proved unable to make much needed savings in costs by wage reductions.

The situation with the numerous English workers, who filled most of the managerial posts and were responsible for all the mines' mechanical equipment, was little better from the companies' point of view. They demanded wages that were three of four times above those paid in contemporary British mines yet also often proved rebellious, insubordinate, and unreliable and on at least one occasion staged a strike. During the companies' declining years, however, they proved less capable than the barrateros in resisting wage reductions. In the early 1840s they submitted to a series of cuts which the company continued until 1845, when the loss of workers to other areas obliged them to reverse the policy.

The companies' administrative, supply and labour problems were compounded by severe technical problems in the extraction and dressing of ore and unwise managerial decisions as to how they might best be solved. In view of his pre-eminence in the managerial structure of both companies, Taylor must bear a large degree of the responsibility in this latter respect. The most probable reason for this uncharacteristic poor judgment was his lack of familiarity with the mines and an inflexibility of mind which prevented him from deviating from his previously successful formula. This is well illustrated in the measures taken to deal with the severe flooding problems at the Real del Monte mines. Taylor had previously several times achieved almost immediate success in reducing costs and restoring profitability at his English mines by abandoning work on the traditional long, expensive and time-consuming drainage adits and installing in their place powerful water or steam powered pumping equipment. Perhaps intoxicated with this success and certainly committed to carrying the wonders of modern British mining technology to primitive Mexican mining districts, Taylor struggled throughout the years of the company's activity at the mines to emulate his domestic experience.

Driving of the deep Aviadero drainage adit, first started in 1816, was reluctantly continued as the Real leases stipulated but with little sustained effort. The main energies were devoted to the installation of the latest steam-pumping machinery, manufactured in Cornwall and transported in pieces over the long and expensive sea and land transport routes for re-assembly at the mines. Unfortunately the real scale of the drainage problems was persistently underestimated and a whole series of progressively larger engines was called for. This culminated in the dispatch of a 70 inch engine just before the closure of the mines in the 1840s, which was as large and powerful as the best then in use in the deep Cornish mines. These problems not only duplicated several times over the already high cost of installing effective steam engines, but in the later years also raised major problems about the availability and cost of essential fuel supplies. Taylor's policy of favouring mechanical pumping arrangements over natural drainage projects may have offered important short-term advantages in terms of costs and immediacy, but in the long-term there was the certainty that pumping costs would continue and perhaps increase while a drainage adit, when finally completed, could provide 'free' drainage for thousands of gallons per hour. In England and Wales chance appears to have justified Taylor's choice of the short over the long-term project. In Mexico, however, it went the other way and there the penalties were even greater.

When the Mexican mining ventures were first launched in 1824 they were received with great enthusiasm by the London financial market as one of the most potentially profitable projects for generations. They stimulated a rush for shares parallelled only by that of the South Sea enterprises more than a hundred years before. Although he was intimately involved in the projection of three major companies and had much to gain from such an atmosphere, Taylor was one of the few who tried to take the heat out of the situation and to restore a more considered and balanced view. To this end he wrote letters to the *Times*[6] and published a short monograph suggesting that the profits from mining in Mexico would probably be no larger than those already obtained from carefully managed British mines.[7] However, not even he imagined that the Mexican bubble would prove as unfounded as its South Seas predecessor and permitted himself to be carried along with the general flood of optimism.

During its first three years of operations the Real del Monte Company, largely under Taylor's direct control, invested a massive £½ million in their Mexican leases. At first its shares commanded a high and buoyant price on the market but as returns proved slow to materialise and losses mounted, they gradually began to slide. Undismayed by the unusually slow progress of the mines, loans were arranged in 1827 and 1828 to help keep up investment but although output gradually increased it continued to fall short of total outlays. By the end of the 1820s the market had become disillusioned with the general prospects of Mexican mining and the company experienced its first financial crisis; a most unusual experience for any company with which Taylor had so far been connected. The failure of the mines to return dividends and recent reports of political instability and violence in Mexico made the stockholders reluctant to provide further funds and Taylor and the directors were obliged to promise greater future economy at the mines and annual investment limits in return for a further call on shares. Under these more restricted conditions operations continued but it was not until 1837, after an expenditure of nearly £2 million, that receipts were finally increased sufficiently to return a small profit. Share values had by this time fallen to less than 3 per cent of their level in 1825 and were little affected. The following year profitability was sustained notwithstanding a sharp increase in the level of expenditure, but thereafter it again disappeared as receipts failed to cover costs which had risen to levels far above anything since 1828.

A second financial crisis for the company ensued in 1839-40, which stimulated a revived search for economy in expenditure. Although there appears to have been very little surplus left to pare off and total expenditure continued almost undiminished, a fortunate upturn in receipts in 1842 and 1843 restored marginal profitability. This again proved short-lived, however, and by the spring of 1845, after yet another year of heavy losses, many of the shareholders were in revolt against the company's management. This was an almost unique experience in Taylor's career, with demands for his removal from office and the transfer of the company's headquarters from his chambers to their own rooms. Although these criticisms were successfully fended off, confidence in the mines had now virtually disappeared and only a miraculous immediate improvement in their condition could sustain further working. This was not forthcoming and the outbreak of the Mexican-American War in 1847 with its consequent disruption of communications proved the final straw. By the beginning of 1848 the company's financial problems became more critical than ever before and in a final attempt to revive confidence John resigned from his protected position as manager[8] and stood instead for election to one of the three vacant posts in the Court of Directors, declaring his intention 'to struggle to the last'. Notwithstanding his past failure to bring the success so badly wanted and paid for by the shareholders, his indisputable eminence as one of the country's foremost mining experts ensured his success in the election. Struggle though he might, however, it was impossible markedly to reduce the now enormously high operating costs at the mines and on 30 October of the same year a meeting of the stockholders voted unanimously to wind up the company.

Although quite high levels of output had been achieved at the mines, profitability had persistently eluded them and well over £1 million had been lost without trace. Ironically, however, Taylor was to be proven right in the long-term. Less than twenty years later the Real mines were taken up by another British company and building on the massive investments of the earlier ventures, the long anticipated rich lodes were finally discovered and returned handsome profits.[9] It was always a fundamental tenet of John's approach to mining that if you have leases of promising ground, you only have to keep your nerve and go on long enough to make them pay.

The winding up of the Real del Monte company was closely followed by the dissolution of Taylor's other, slightly less abortive enterprise at Bolanos. These were the last of the numerous Mexican mining ventures launched in the feverish years of the mid-1820s and were the only major blemish on his otherwise successful business career. The problems confronting these enterprises had been generally no different in type from those encountered by ventures in the more remote mining districts of England and Wales but they were infinitely greater in scale and required a very different approach. It was essentially this that John and his fellow projectors failed to grasp. When that ever present and important element in mining, chance/luck/fortune, whatever you call it, also tipped against the enterprises, there was nothing that even Taylor's great talents and experience could do to ward off disaster. As R. W. Randall, the historian of the Real del Monte mining company, so appropriately entitled his work, it was a study in the *Anatomy of Failure.*[10]

REFERENCES

1 J. Taylor, *Statements Respecting the Profits of Mining in England Considered in Relation to the Prospects of Mining in Mexico* (1825), p7.

2 J. Taylor (ed), *Selections from the Works of Baron de Humboldt, Relating to the Climate, Inhabitants, Productions and Mines of Mexico* (1824).

3 This large single-share value was characteristic of the cost-book companies which John customarily formed to operate his British mines.

4 *Law Journal* (Report) March 1825, p68f.

5 G. Henwood, *Cornwall's Mines and Miners* (Truro, 1972), pp222-4.

6 22 Nov 1824, p3.

7 *Statements Respecting the Profits of Mining in England Considered in Relation to the Prospects of Mining in Mexico* (1825).

8 Taylor's appointment as manager was written into the company's deed of settlement and he could not legally be removed.

9 *MJ*, 11 April 1863, p251.

10 R. W. Randall, 'The Anatomy of a Failure: the British Real del Monte Mining Company in Mexico 1824-1841' (PhD thesis, Harvard University, 1965) now published as *Real del Monte: A British Mining Venture in Mexico* (University of Texas, 1972). I am most grateful to Professor Randall for permission to draw heavily on his work in this section.

CHAPTER 4

Consolidation and Expansion

Although he suffered a set-back in the Mexican ventures, John Taylor was certainly not alienated from all further foreign mining projects. Throughout the rest of his life he and his sons took an interest in or acted for several such enterprises, including the Nouveau Mond Gold Mining Company, working mines in California and Spain in the early 1850s and the Linares Lead Mining Company in Spain.[1] At home they also continued to expand their operations by taking up leases, forming companies and acquiring consultancies in both their existing areas of operations and in new districts.

I Cornwall

In Cornwall they expanded both their copper and tin mining interests by taking up more mines and consolidating their operations in several districts. In the St Austell district, for example, activity and production at the Polgooth, Pembroke, Crinnis, and Charlestown United mines, which were first acquired in the early 1820s and before, was steadily increased by heavy investment in underground and surface operations.[2] In 1834 Taylor also set up his own tin-smelting works at Charlestown but ore supplies proved inadequate and the works were closed at the end of the following year.[3] In the Gwennap-St Day district they took on the large United Mines complex in the late 1830s to compensate for the loss of their Consolidated leases, while in the Redruth area they acquired interests in Wheal Spavnar, Pednandrea, Wheal Busy and the Tolgus mines.[4] Other mines near Hayle, Camborne and St Agnes were also added to their list, including Polberro Consols, taken up in 1848 on extremely favourable royalty terms, largely in order to reduce unemployment in the St Agnes district. By this date, Taylor and his sons had also become mineral agents to the Duchy of Cornwall and may have felt some obligation to practice a little philanthropy. With such extensive interests in the county, John was an obvious choice to represent the Cornish mining interest in a petition to Parliament in the early 1840s against further reductions in the copper and tin tariff.

II Central Wales

Although Cornwall remained the largest and most important area of the family's activities, the most significant new departure of the period was a move into the Central Wales lead-mining district of Cardiganshire and Montgomeryshire. In 1824 the Williams family of Scorrier, Taylor's great rivals in Cornwall, had leased Lord Lisburne's mines and twelve others on the Gogerddan estate in this district. They resurrected them and worked them to some profit during that decade, using techniques similar to those of the Taylors, but times were not good in the metal market and a continued slide of prices soon eliminated all profits. By 1832 they had lost over £20,000 in the venture. The following year they surrendered all but one of their leases, that of the Llanfair Silver Mine.

The Taylors, whose interests in the Flintshire/Denbighshire district were facing similar, but as yet less serious problems, took a more optimistic view of the long term prospects of the district. John Taylor, who already had some experience of the area from a small interest in the Alderson's Cwmystwyth venture, decided to extend the family's interests westwards and in May 1834 took over the unexpired part of the Williams' Lisburne lease and the company's equipment for a payment of £5,000.. In their now customary manner, the Taylors formed a cost-book company, known as the Lisburne Mines Company, in which they were the principle shareholders, to work the property, and entrusted its day-to-day management to Matthew Francis, nephew of Absolem Francis, the manager of their Halkyn mines.

Proceeding cautiously at first and confining most activity to the neglected Frongoch Mine, in little more than a year they had restored the workings to profitability. The pattern was much the same as in their other extensive properties; intensive investment in carefully planned underground exploration and detailed attention to the layout and mechanisation of the dressing floors. To maximise on the use of limited local water-power resources, new leats and reservoirs were constructed and machinery, made at their own foundry at Rhydymwyn near Mold, Coed Talon ironworks near Wrexham, or purchased secondhand in Cornwall, were imported coastally through Aberystwyth.[5] By the end of the decade the other major Lisburne mines, Logaulas, Glogfach and Glogfawr, were also progressing satisfactorily and in nearly sixty years of continuous working they, together with Frongoch, are estimated to have produced 107,174 tons of lead ore, 50,000 tons of zinc blende and a little copper, with a total value of nearly £1½

12 The ruins of the engine house for a 60in steam pumping engine at Frongoch Mine, Cardiganshire, in 1974. *(R. H. Bird)*.

million.[6] For an initial investment of £7,500 the Taylors and their partners divided dividends of nearly £1¼ million.[7] On top of Williams' failure in the footsteps of the Taylors at Consolidated, this story of the Taylors' success at their former mines must have been a bitter pill indeed. It perhaps says much for the relative mining abilities of the two family enterprises.

The success of the Lisburne mines encouraged the Taylors to look for further leases in the district and again like the Williams before them, they took up mines on the neighbouring Gogerddan estate. As early as 1836 they revived the old Goginan workings and within sixty years had possession of all the major mines on the estate. At first they met with only mixed success. Goginan proved extremely productive but the nearby Darren and

Cwmsymlog mines swallowed £14,000 in underground exploration costs without finding worthwhile deposits. Nevertheless, the general success of the Taylor mines, together with sustained good lead prices, was sufficient to start a 'rush' in the area and many small partnerships of local tradesmen were set up to prosecute small marginal workings.[8] These years also saw changes at the established Lisburne mines. Matthew Francis, who had performed so well as manager of the mines during their early years, appears to have become progressingly incompetant and unsatisfactory and in 1842 was replaced by John Taylor Jnr, who already had control of the North Wales mines, assisted by George Fossett.[9]

By the end of the decade the Taylors had sufficient confidence to contemplate still further expansion and in 1848 took up all the major leases on the Nanteos estate. This included the old Cwmystwyth Mine which had been worked for the last three years, until bankruptcy, by a company under the management of Matthew Francis, the ex-manager of the Taylor's Lisburne leases. By wisely building on their predecessors' unsuccessful

13 Plan of the dressing floors at Goginan Mine, Cardiganshire. *(Memoirs of the Geological Survey*, 1848)

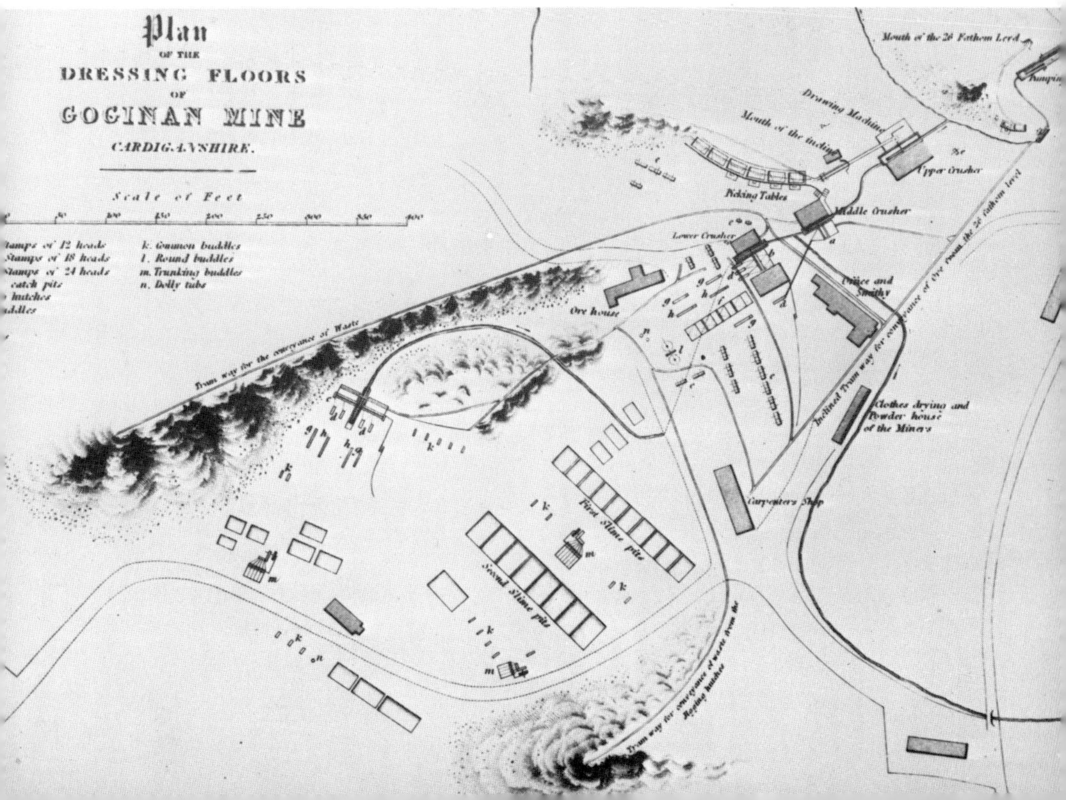

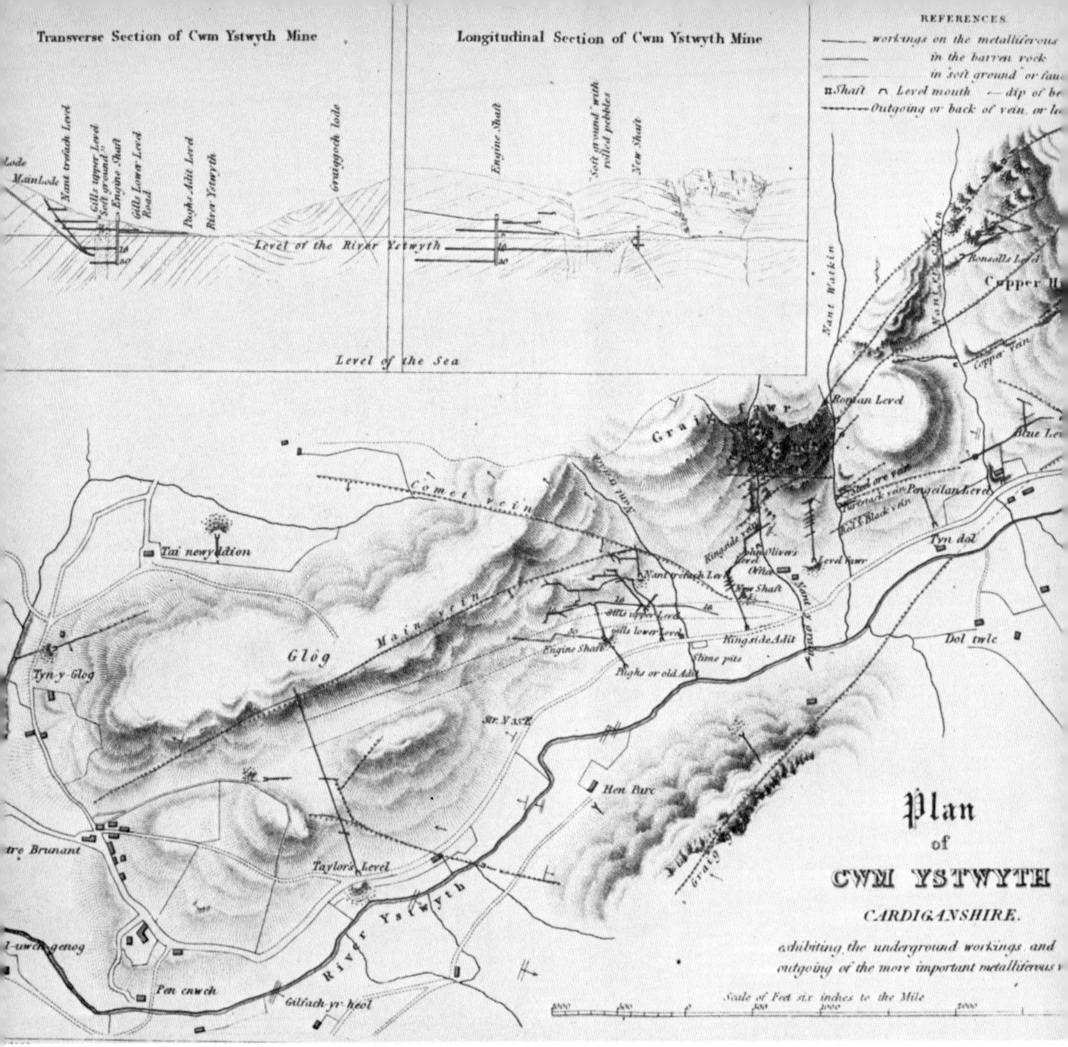

14 Plan of Cwmystwyth Mine, Cardiganshire. *(Memoirs of the Geological survey*, 1848).

investment, the Taylors again succeeded in restoring the moribund mines to profitability and initiated a new era in the development of the estate. Within two years shares in the Cwmystwyth venture stood at 50 per cent above their original price and within ten years, its shallow, complex workings were among the most productive in the county. In 1857, Cwmystwyth together with the Taylors' Frongoch, Logaulas, and Glogfach mines on the Lisburne estate, and East Darren on the Gogerddan estate, produced over 70 per cent of Cardiganshire's total lead output, with other important percentages being accounted for by the Taylors' smaller and sometimes even more profitable ventures, such as Goginan. Most of these mines continued as productive and profitable ventures into the last quarter of the nineteenth century, long after John Taylor's own active involvement was over.

Nearly all of the Taylors' mines in Central Wales, like those in Devon and Cornwall, were worked by companies formed on the cost book principle, with the Taylors holding a large and often controlling interest. Officially this method of organisation, which was a form of large extended partnership, was illegal outside of the jurisdiction of the stannary courts of the South-West where it originally had been evolved, and Derbyshire where it was regulated by ancient laws and customs to a considerable degree. Taylor, however, like most Cornish adventurers in Wales, accepted these risks (which in practice were minimal) because of his long experience and familiarity with that system and, more importantly, the advantages which it offered for the close control of policy-making and managerial activity. The significance of this latter factor is illustrated by the average size of the cost-book companies which the Taylors controlled. By the mid-nineteenth century it was common for many of the larger cost-book mines, where capital requirements were extensive, to be divided into many thousands of shares, sometimes more than six thousand. However, wherever possible the Taylors endeavoured to restrict the number of their shareholders to between 64 and 112. This required very large individual subscriptions, but ensured much greater ease of communication and agreement within the enterprise and a high degree of control over who purchased shares and therefore the degree to which the company could be penetrated by unscrupulous speculators.

The cost-book system did, however, have one serious drawback; it allowed for no continuing stock or financial reserve. At the end of every period of account, usually three month, all profits or costs were equally divided among the shareholders and a clean sheet drawn up for the next period. This could obviously present serious problems for long-term development projects and carefully planned working and was clearly an important contributory factor to the wasteful interrupted pattern of working at many mines. For example, a group of adventurers would subscribe the capital for exploration and development over several accounts and if an ore body was discovered it would be quickly looted and all the profits equally divided. Finance might then be subscribed for further exploration but if no new lodes were quickly found the company would dissolve before, or perhaps after, the profits that had been made so far were dissipated. The mine might then remain idle – being allowed to flood and the machinery to deteriorate or be sold off – before another group found the confidence for

15 A model of the hydraulic-pressure pumping engine erected at Guy Shaft on the Alport Mines, Derbyshire in 1842. This engine was fixed 210ft below the surface and had a 50in diameter single-acting cylinder. *(Science Museum, London).*

further enterprise and went to the expense of restoring the original workings.

Taking the industry as a whole, this inevitably caused great waste and inefficiency over time. The Taylors successfully endeavoured to overcome this problem and ensure proper long-term investment and development by adapting the cost-book system to allow for a continuing stock. In their case this was not held in the form of a financial float but often in broken ore which was stored underground; a large continuing stock which could be drawn on and cashed as required. This was perhaps one of John Taylor's most important contributions to mining management and a major factor accounting for his sustained success.[10]

III Derbyshire

At the same time as John and his sons were expanding their Central Wales interests, they also moved into the ancient lead-mining area of the Derbyshire Peak. Although now long-past its most productive period, there were still good prospects of large returns in the area and extensive interests were acquired in the Longstone Edge, Hubberdale, Magpie and Alport mines.[11] Of these, the Alport mines were undoubtedly the largest and most important. A cost-book company, known as the Alport Mining Company, was launched at the beginning of 1839, under the direction of John Taylor and his eldest son, to work leases of several mines in a 2 square mile area near Youlgreave, which were currently drained, inadequately, by the Hillcarr Sough and its Thornhill and Stanton branches. These workings had been seriously troubled with water for some years and it may be that Taylor was introduced to the district either by the Duke of Devonshire from his Ecton and Grassington Moor mines or by the Duke of Rutland, the owner of the mineral rights in the manors of Harthill, Haddon and Youlgreave where most of the mines lay, in an attempt to revive production.

The plan, worked out by John Taylor and Captain Eddy, one of his most senior and experienced managers who had carefully inspected the mines, was a typically ambitious one. It was to lower the water in all of the mines through one co-ordinated drainage system, employing a combination of pumps and extensions to the Hillcarr Sough, the latter having the additional advantage of proving new ground. As Nellie Kirkham records in her detailed history of the project, it was probably, 'the only time in the history of the draining of the Derbyshire lead mines when the unwatering of so many mines and veins was planned as a whole.'[12]

The main problem at the mines was working proven deposits of good-quality ore from below the level of the drainage soughs. The nature of the terrain prohibited the driving of deeper drainage levels and the most practical solution was to install pumping engines, either on the surface or underground, to raise the water from the lower workings and to discharge it along the existing soughs. This had been tried before, using waterwheels, steam engines and hydraulic engines, but none were ever powerful enough to cope with the problems in the previously unco-ordinated and piecemeal way. Taylor's plan was in three parts: to reduce seepage of water into the lower workings by clearing and sealing surface and underground drainage channels; to deepen existing shafts below the level of the soughs using hydraulic engines, including a new and powerful 50in double-acting engine; and thirdly, from the profits of working the first ground made available, to install an 80in Cornish steam pumping-engine for still deeper operations. These improvements, which were put under the charge of John Darlington, the brilliant hydraulic engineer who worked for the Taylors in several of their mining districts, were estimated to cost nearly £20,000 and would take at least two or three years to complete. In the event, not all of these plans were implemented, in particular the steam engine was never built.

As well as improvements to the draining of the mines, Taylor also proposed major changes in their management and methods of working. These changes would result from his usual desire to substitute what was generally known as the Cornish system for existing Derbyshire practice. In essence this amounted to the employment of a larger number of managerial staff than was common in Derbyshire. Anticipating opposition from other shareholders for what might have been considered as 'unproductive' expenditure, John Taylor prepared his defence in advance in a report submitted to the proprietors in September 1839. It provides a succinct statement of his approach to the organisation of mining. He declared,

> I must state my strong persuasion that true economy is consulted by employing a sufficient number of experienced agents, and that absolute loss, and *that* to an extent not easily estimated, is incurred by most of them . . . Let it . . . be remembered that the great expense of mining is incurred by the employment of labour; that it becomes therefore of the utmost importance that this labour be properly directed in the first place; and secondly that it be constantly and unremittingly stimulated by the most active and intelligent superintendence. And I may add, that in a long experience I have found no way of accomplishing this object, but by employing a sufficient number of persons, whose training has been such as to qualify them for the due execution of their duties; and I venture to say that the most profitable mines have been those where a very active system of management has been kept up and enforced. I ought to add, that I am convinced that the charges of labour underground, at the Alport Mines, at present, both for breaking the ground and getting the ores, are higher by a very

> large proportion than they ought to be, were there suitable arrangements for vigilant and skilful supervision.
>
> I must remark further that the book keeping on mines is an important branch of the management, and ought to be committed to a person properly trained to the business; and that if accounts for such a concern as the Alport Mines will become, be kept as they ought to be, and the duty of receiving and delivering the stores be attended to in the manner which may best insure their economical application; there is ample employment for a clerk duly qualified and experienced in mining and book-keeping.
>
> I consider further, that it is wrong in principle to allow a mine-agent to keep account. First, because in this manner all check is lost; and any system defective in this respect must be very unsatisfactory. Secondly, because it is an occupation for which the habits and education of mine-agents seldom fit them. And thirdly, if there were but little stress in these objections, it is still clear that either the accounts are neglected or the time and attention of the agent are directed into a course inconsistent with his special duty of vigilant control of the mining operations.[13]

During the first few years of activity at the mines work proceeded reasonably well, but an increasing flow of water into the lower workings as they progressed at depth together with problems with the delivery of machinery meant that profitability eluded the company. Taylor's annual reports to the shareholders expressed a convinced optimism in the final outcome of the venture but he must surely have had nagging doubts that he might again be confronted with the same insoluble problems which at that very moment were bringing the long promising Real del Monte mines to near disaster. However, a steady increase in output helped sustain and even improve confidence and in 1841 a share in the company was worth £435, more than three times its value the year before. Finally, in his report of September 1845, Taylor could happily record that,

> I can congratulate the proprietors upon our meeting under more favourable circumstances than heretofore, and when I can hold out a prospect of a profitable result from our future operations. The statement of accounts submitted to the meeting shows an excess of returns above the costs during the past twelve months of £689 19s 6d, this sum and the amount of a call of £720, together £1,409 19s 6d less the balance against the concern in September last of £425 13s, leaves a balance in hand of £984 6s 6d. The profitable working of the mines, however, has but lately commenced, and the drainage of the ore ground below the eight fathom levels (measured from below adit), from which the main returns must be drawn, is still but partial; every week may now be expected to bring more ground into return.[14]

A dividend of £10 for every £100 share was declared, payable before the end of the month. Taylor's optimism about the future was more than justified the following year with a further £2,880 distributed in dividends, the settlement of outstanding debts, and the carrying of a net balance to the next account of £942 8s. In his report of September 1846, however, Taylor made reference to a problem which was to plague the company through the rest of its existence and was at least partly responsible for the final cessation of activity in the mid-1850s. The financial position for the year, he records,

> would have shewn a greater amount of profit had not considerable interruption been met with in many parts of the mines, from a deficiency of surface water for the [hydraulic] engines, and that for a much longer period than for many years before. The want of control over the supplies of water afforded by the Lathkill and Bradford rivers is a serious detriment in the interests of this concern. The dams upon their courses are frequently drawn off at the very driest season of the year, and the sudden excess of water flows away to waste, and much time is afterwards required to replenish the emptied ponds. Not only is any regularity of return from the mines placed beyond our power of computation, but a great number of workmen are thrown out of employ at the very season of the year most favourable for their operations.

Most uncharacteristically for a man of such immense experience and resourcefulness, and an ominous indication of the scale of the problem, Taylor went on to declare pessimistically that,

> I am at a loss to suggest a remedy for this great drawback to the most advantageous working of these mines, knowing the difficulties which attend the management of the mines, but at the same time I must state that if any arrangement could be made to secure the full and free use of the water power afforded by them, it would be followed by very beneficial results to all the parties interested.

(The latter remark relates to their current obligation to leave sufficient water in the rivers for other users, such as mills, farmers, fishmen etc.)

In the account to September 1847 profits continued to be made but were considerably less than had been expected. Adopting a cautious approach, the total amount distributed in dividends (NB dividend payments and calls in cost-book companies such as this were made every few months – in this case monthly) was reduced to £720 and the balance kept in hand against unforeseen changes in circumstances was increased to £1,639 6s 9d. Taylor's report recorded that some of the lodes had proved less productive than anticipated and that,

> the very frequent suspension of our underground operations by the continuous want of a sufficient supply of water . . . causes a complete stoppage to our returns for a time, whilst the charge of our establishment must remain the same, and increased wages be allowed to the men when again employed, in order to compensate them in some measure for their loss of labour from such causes.

Perhaps with a view to squeezing concessions from the Duke of Rutland for the more liberal use of available surface water supplies, he continued pessimistically:

> This interruption to our workings, by a want of water for our engines, has so seriously affected us during the past two or three months, that I feel convinced we cannot hope to carry on the Mines with the present mode of drainage, *under such restrictions,* either with profit to ourselves or advantage to the neighbourhood.[15]

The water flowing into the mines, he declared, was between 2,000 and 6,000 gallons per minute; the largest on record in the history of mining. This latter claim might not have been entirely accurate but pumping such quantities during the dry season by steam power would undoubtedly have required an investment and running charges far beyond the financial capabilities of the mines. The only practicable solution to the drainage problems was the use of hydraulic engines, and these had proved admirable for the purpose, but they must be ensured adequate 'fuel'. The only immediate solution was to accept the inevitable and to reduce the depth to which the mines were pumped in dry weather. Taylor noted that, 'By the adoption of this plan . . . a portion only of the water in the rivers would be required, and thus a considerable number of workers might obtain employment on ground that might be kept in reserve with a view to this object.'

Notwithstanding these arrangements, the next period of annual account, to September 1848, saw a large overall loss of £1,214 1s 1d. This was partly the result of a slight fall in production but more particularly was due to heavy expenses incurred in the purchase and installation of yet another large hydraulic engine and a severe fall in the price of lead ore, which Taylor estimated as making a difference of £1,450 to the company's receipts calculated on the basis of average prices the year before. The following account, to September 1849, produced little relief. A total loss of £627 4s 11d was made over the year and it had been necessary to make a call on shares of £720, the first for several years, to cover costs and maintain a small balance in hand. An unusually long summer drought, which had already lasted three months and was still continuing when John presented his annual report, had caused a complete suspension of working below Hillcar Sough and had seriously reduced output. A rigourous economy programme was put into operation but failed to compensate for the sharp drop in receipts which had been aggravated by a further fall in the price of lead ore. Taylor wrote alarmingly, 'The price of our lead ores throughout the year has averaged only about £9 per ton, which is a rate below that at which such mines as these

16 12 Bedford Row, John Taylor's London home during the 1820s. It was conveniently situated for his offices in the city at 14 Chatham Place, Blackfriars.

17 31 Chester Terrace, Regent's Park, John Taylor's home for the last ten years of his life.

can produce it advantageously.'[16]

Caught between the nutcrackers of immense technical problems and high operating costs on the one hand and declining ore yields and stagnant prices on the other, the company struggled on for a few more years assisted by the remittance of all royalty payments by the Duke who was anxious to keep his property in work. Finally, however, in 1852 the decision was taken to wind up the company and to advertise its machinery for sale. Some of the proprietors continued in a new company to work the small deposits above sough level but even this activity had ceased by the end of the decade. The Alport venture had been a bold and promising project defeated by problems beyond the capacity of contemporary technology. It was one of the still few major failures of John Taylor's now lengthy career. It bears some similarities to the disappointment experienced at the Real during the same years, but the overall scale of the financial losses was considerably less.

IV Other Areas

As well as their extensive interests in Cornwall, North and Central Wales and Derbyshire, John Taylor and his sons also became involved in more localised, but nevertheless important enterprises, in most of the country's other major mining fields. In the mid-1830s, for example, they took up leases to work mines on Lord Dungally's Tiperary estates and in County Clare[17] in Ireland and formed a company to operate the rich Coniston Copper Mine in the Lake District. They were treated to the usual Taylor formula of heavy planned investment, technological improvement and managerial innovation and produced substantial improvements in output and profits. The Coniston Mine in particular, under the management of Captain John Barratt, one of their most long-serving and experienced managers, passed into a new era of prosperity. In 1849, when employing 400 men and yielding 250 tons of ore per month, it ranked as one of the largest and most productive copper producers in the county, supporting the construction of its own mineral line to the coast at Broughton.[18] By the early 1840s the Taylors were also acting as advisers to the Commissioners of Woods and Forests with regard to their mineral property in various parts of England, Wales, and the Isle of Man. In the 1840s, for example, Richard Taylor advised on the leasing of the Rialton Estate in Cornwall and in the 1850s, Captain Stephen Eddy, their manager at the Duke of Devonshire's Grassington mines, arbitrated in a request for a reduction of mineral royalties at the Grinton lead

18 Coed Du, the North Wales home of John Taylor from 1823 and later his eldest son. *(C. J. Williams).*

mines in Yorkshire.[19] Although John was now well into his seventies and progressively less interested in the day-to-day affairs of the business his earlier expansionist policies were carried on by his sons. They continued to expand their direct involvment in mining by taking up leases and forming companies and took on new agencies, acting for the Marquis of Bath from 1861 in the exploitation of his rich lead property in the Snailbeach area of Shropshire[20] and the Earl of Derby from 1864 in the development of his North Pennine property.

V Organisation of the Company

In nearly sixty years of active working life John Taylor assembled a business empire which matched in scale and geographical dispersion the largest of its day in any branch of industry and commerce. Its growth was made possible and it continued to operate efficiently, only by the parallel evolution of a complex web of managerial command which left John Taylor free at the centre to receive reports and formulate policy and delegate responsibility and authority for everyday affairs to a proven team of some of the country's most qualified mine captains, such as the Eddys, Barratts and Francises. John's two sons, John Jnr and Richard, also came to play a key role in this system. After a practical apprenticeship, during which they spent time at mines in Britain as well as Hungary and elsewhere on the continent,[21] they were introduced into the business from the late 1820s and gradually evolved a regional pattern of responsibility, becoming partners with their father in 1837. John Snr worked from what were to become the partners' London head offices at 14 Chatham Place, Blackfriars, [22] which were convenient for his financial contacts in the City, while his eldest son, John Jnr took charge of the Welsh and North of England operations from Coed Du, the family's home near Mold in Flintshire, and Richard co-ordinated their interests in Cornwall and the South-West where he also acted for the Duchy of Cornwall.[23] During the eighteen thirties and forties John continued to exercise a strong direct control over the company's affairs by constantly travelling between the many mines and districts under his charge but by the eighteen fifties, after several prolonged illnesses and now into his seventies, he virtually gave up these activities and began to lean increasingly on his middle-aged sons.

John Taylor Jnr's evidence to the *Select Committee of Enquiry into the Rating of Mines*, published in 1856, gives some insight into the organisation of the business in John's later years. Describing himself as a Civil Engineer representing a large mining interest throughout England and Wales, he declared that his 'own special districts are in Cardiganshire' though he spoke with first-hand authority of mines in Flintshire, where he had lived and acted as a Poor Law Guardian for many years, and of mining areas in the Pennines, from Derbyshire to Northumberland, which he visited frequently and some of which he had charge.[24] In reply to questions about Cornwall, he replied, 'My own particular business has been more out of Cornwall than in it. Still, Cornwall is a country which is perfectly well known to me.' The management of an important tin mine near St Agnes gave him a continuing foothold in the area and a constant flow of communications with his brother, Richard, kept him informed of the general developments in the district.[25] In contrast to specific questions answered from his own knowledge and experience, a question about general policy, such as whether or not metal mines should continue to be exempt from the poor rate, brought forth the reply, 'speaking for my father, and my brother, and myself, who consult together upon these matters, the opinion at which we have arrived is, that the exemption from rating is just as to mines worked upon veins'.[26] When John Snr finally retired from the business in 1861 at the beginning of his last long illness, it was continued by John Jnr and his brother under the now long-established and well-known name of John Taylor and Sons. They continued to expand some operations and contract others to keep the business abreast of changing conditions at home and abroad but the essential policies and procedures developed and proved in their father's day remained the same. There is perhaps no better tribute to the validity of these policies than that they were a continual success for the family firm through another three generations, until they finally closed their doors in 1969.

REFERENCES

1 *MJ*, 14 Nov 1854, p754 and 24 Feb 1855, p116.

2 See D. B. Barton, *A Historical Survey of the Mines and Mineral Railways of East Cornwall and West Devon* (Truro, 1964), pp9-18. In the mid-1820s copper output at Pembroke was second only to Taylor's own Consolidated Mines.

3 D. B. Barton, *A History of Tin Mining and Smelting in Cornwall* (Truro, 1967), p63.

4 D. B. Barton, *Copper Mining.*

5 Waterwheels for the Frongoch workings were purchased in 1841 in Cornwall when Wheal Jane and Wheal Gwallon sold up. An experiment in that same year with a steam winding and crushing engine at the mine to supplement water power was abandoned in 1846 due to the high cost of transporting coal fuel to the district. See D. E. Bick, *The Old Metal Mines of Mid-Wales,* 'Pt 1 Cardiganshire – South of Devils Bridge' (Newent, Glos, 1974), pp13, 17.

6 W. J. Lewis, *Lead Mining in Wales,* p175.

7 D. E. Bick, *Metal Mines,* p11.

8 W. J. Lewis, *Mining in Wales,* p176.

9 D. E. Bick, *Metal Mines,* p24.

10 D. B. Barton, *Copper Mining,* p74.

11 N. Kirkham, *Derbyshire Lead Mining Through the Centuries* (Truro, 1968), p115.

12 N. Kirkham, 'The Draining of the Alport Lead Mines, Derbyshire' *Trans Newcomen Soc,* **XXXIII** (1960-1), p79.

13 *Mr Taylor's Report Respecting Deeper Drainage of the Alport Mines : To the Proprietors of the Alport Consolidated Mines,* 23 September 1839, Derbyshire Record Office.

14 *Alport Mines Report,* 11 Sept 1845, Derbys RO.

15 *Ibid,* 9 September 1847.

16 *Ibid,* 20 September 1849.

17 G. A. J. Cole, *Irish Mineral Resources* (1922).

18 W. G. Collingwood, *The Book of Coniston* (Kendall, 1906), p60.

19 Report of the Select Committee on the Rating of Mines, *BPP* (1856) **XVI** (evidence), p80 and PRO, Crown Estate Records: Correspondence Papers; Crest 2 Nos 174 & 1391.

20 Wilts Record Office, Acc 845; F. Brook and M. Allbutt, *The Shropshire Lead Mines* (Moorland Publishing Co, 1973), p24.

21 In a letter to the Secretary of University College London 3 April 1828, John refers to his sons' current visit to the mines of Hungary and their conversations with the eminent natural philosopher, Dr Tipser of Neusohl. See UCL College correspondence, D.M.S. Watson Library.

22 These offices were leased by John on the 20 April 1825. Redevelopment of the area following the construction of Southwark Bridge changed this address to 6 Queen St Place and future generations of partners retained their offices there until they were destroyed during World War II.

23 National Library of Wales, Album of Richard Rees, Alltycham Pontardawe, MS 11138D. Letter from John to Richard Rees dated 5 August 1848.

24 He appears for example to have taken over from his father in advising the Greenwich Hospital on the working of their Alston Moor property, acted for Mr Beaumont as an arbitrator in a dispute concerning his Weardale mines, and to have had control of their own mines near Bakewell in Derbyshire. See *BPP* (1856) **XVI** (Evidence), pp1122, 1251, 1253.

25 *Ibid,* 1122.

26 *Ibid,* p1119.

CHAPTER 5

Social and Scientific Interests

I The Geological Society of London

The increasing managerial assistance rendered to John Taylor by his sons and staff of paid captains enabled him not only to expand and keep control of his business affairs but also permitted him time to pursue a wide range of intellectual, cultural and political interests. As suggested earlier, John's family background had strongly emphasised the importance of such pursuits from early childhood and in his later life he was to become one of the most influential educationalists and men of science of the mid-nineteenth century. Within a year of arriving at Wheal Friendship, when heavily engaged in establishing his career and reputation, Taylor had found time to publish his first article, 'A sketch of the history of mining in Devon and Cornwall', published in the *Philosophical Magazine*[1] and by 1807 had established a sufficiently impressive reputation as a geologist and mining engineer to successfully seek election to the newly-formed Geological Society of London. Delighted with the platform which this society provided for scientific and practical discussion he later published several articles in their transactions. Once resident in London he became an active member of their committee, treasurer from 1816 to 1847 and for many years later in life a vice-president. From such a position he became well acquainted with many of the founders of modern geological science, such as Henry de la Beche, R. I. Murchison, and W. J. Henwood and undoubtedly played a significant role as a link between the 'scientific' and 'practical', helping to produce the interaction which was so important for the continued progress of science. This was a role that he was to play time and again in many different spheres.[2]

II Miners' Education

Shortly after leaving Devon, John Taylor's help was enlisted for a project to establish a small elementary school in Tavistock. It marked the beginning of what was to become an active lifelong interest in the improvement of educational facilities at all levels. Intellectually these activities related closely to his interest in managerial systems and the desire to maximise the efficiency of men in the same way as maximising the efficiency of machinery, through qualitative improvements. The Tavistock project also sheds light on John's attitude to the religious controversies which bedeviled the development of general education for much of the nineteenth century.

When still resident in Tavistock, Taylor had exhibited a certain community spirit by his buildings in Taylor's Square for offices and other facilities for the mining interest and by taking a hand in the foundation of the Tavistock Library.[3] Local residents endeavoured to widen this interest and to employ Taylor's influence with local dissenters to settle a dispute which was currently frustrating the establishment of their school. The tone of his letters clearly illustrates his own position in the matter and shows that at least this leading member of the Unitarian Church did not regard himself as part of the broad ranks of Dissenters. Decrying their activities, he declared that, 'you may find more of bigotry and self-sufficiency among the old stiff dissenters than many other classes of Christian in England. They have the fanatacism and ill-breeding of the Puritans, without their independent spirit and love of freedom'.

John's own solution to the argument about how religion should be taught in the school prefaced that adopted nearly a half century later in the Education Act of 1870. He advocated, 'a perfect separation of all religious matters, because they must and will give rise to dispute and offence and because I do not wish to see religion taught by authority – but as I value education in practice, more than opinions of this sort in speculation, I would, aid in any plan that was not repugnant to real religious liberty.'[4] John lent such moderating influence as he had, in letters to Mr Evans, the leader of the local non-conformists, but no immediate break in the deadlock appears to have been achieved.

Some years later, in a *Prospectus of a School of Mines in Cornwall,* published in 1825 (included in his book *Records of Mining* four years later) John Taylor carried his crusade for better and more general education a stage further. This time, however, it was directed not at local elementary instruction but at the need for a School of Mines. This was to be an institution which would provide the intelligent working miner in Britain with the same level of advanced specialist instruction in the theory and practice of mining as was then only to be obtained in the great continental mining academies, such as Freiberg and Schemnitz. Again Taylor emerged as one of a still small group of early pioneers who already favoured a challenge to Britain's technological and scientific education. Unfortunately it was not for another half century, by which time it was perhaps already too late, that this warning even began to be generally accepted and acted upon by the leaders of British industry.

John's *Prospectus* set out the arguments for a School of Mines. He argued that recent years had seen great progress in a wide range of scientific and technological arts relating to mining and noticed that all this had taken place in England with neither institutions of public instruction nor books. Although he clearly agreed that 'the practical management of mines must always be deputed to those who have gained a most important part of their experience by actual work underground' (perhaps tongue-in-cheek considering his own early background), he also insisted that all captains should be informed about the latest theoretical and practical improvements in their field so they might fulfill their office as efficiently as possible. He observed that once miners were limited in the scale and productiveness of their working but that now,

> ingenuity and science have removed the barrier from those who are placed in the way of instruction and comparison, but to many, who have not these advantages, the door of improvement yet remains in a manner closed; they hear indistinctly of the progress of others, but, not having the means of judging rightly, they are too apt to fence themselves in with prejudice and conceit, and to remit what has not originated with themselves.

He noted that:

> the education desirable for a miner is then a peculiar one, and must be adapted to go with the necessary labour, and not to supersede it; it should explain and make clear the reasons for each proceeding, not make the scholar unfit for his proper duty; it should not tend to the paths of theory and dispute, but show that good practices depend upon solid and intelligable principle.

What was needed was a School of Mining,

> where young men, while they could continue their practice underground, might hear those sciences explained which would be most useful to them, and might devote some of the spare hours which a miner's life affords, to seeing and comparing the practice of others in a place where their art has reached the greatest perfection.

To counter the common suggestion that the average miner was not capable of such instruction, he wrote,

> with regard to the capacity of the working class of miners for instruction but little requires to be said; it is, I conceive, admitted by those who know them best, that they possess it in a peculiar degree. From this class the agents even of the largest mines are taken, and if I were to say much of what I think of the talents that they commonly possess, and of the excellent use they make of the means of instruction, slight as they are, which are thrown in their way, it might appear

> that I meant to flatter men with whom I am much associated and to whom I am so much indebted, but the fact is notorious, in Cornwall particularly, that education is much sought after among the miners, and that its benefits are improved as much as the means will admit, and even far beyond what could be expected from the few opportunities at present afforded them. Miners in general are a superior class of men, and, in the deep mines particularly, the constant exercise of judgement and thought which is necessary, produces a proportionate degree of intelligence.

Concluding with the succinct observation that, 'the demand for intelligent and well-instructed miners is now greater than in any former period' and that, 'a School of Mines will not only be the means of instructing such agents, but it will be a place where character will be developed, and as it were put upon record' – highly important consideration in an industry plagued with rogues and swindlers – Taylor went on to itemize the detailed financial and teaching arrangements for his proposed school.

Intense pressure of work, competing interests elsewhere and the general ambivalence with which the mining interest as a whole received the idea of a Mining School, meant that Taylor made little real progress. Nevertheless the points which he made were well taken by several of his influential contemporaries. The arguments of the *Prospectus* – particularly the need for the combination of practical experience with theoretical instruction – were taken up by Sir Charles Lemon and Robert Hunt from the late 1830s and culminated in the foundation of The Miners' Association of Cornwall and Devonshire in 1859, from which later evolved the Camborne School of Mines. Some credit must also be given to John's early efforts for the foundation of the Museum and School of Mines and Science Applied to the Arts in 1851, which was later to become the Royal School of Mines.[5]

III University College London

During the latter part of the 1820s, John Taylor's spare time was largely taken up with the affairs of the newly formed University College of London. Its non-sectarian 'godless' principles appealed greatly to the educational ideals which Taylor had long since declared in the controversy over the Tavistock School and he became a founder member/proprieter in 1826. From 1828, when the college opened, through the 1830s he served on the council of the college, taking a hand in the formulation of policy and standing reference for various staff appointments, including that of C. B. Vignoles to the country's first chair of civil engineering. Pressure of business and long trips away from London frequently prevented him from visiting the college, however, and finally in 1841 he was excluded from the Council for poor attendance.[6] Although he does not appear to have been much aggrieved by this, it strengthened his resolve to devote more time to the college and in that same year he stood for the post of treasurer. He was narrowly defeated in the elections by William Tooke but stood again the following year and this time was successful. Thereafter he was re-elected every year until at the Annual General Meeting of Proprietors in February 1860 he declined to stand again.

During these eighteen years as treasurer he placed the financies of the college on a sound and continuing footing and as chairman of the Committee of Management exercised great influence over every aspect of the college's activities, including the appointment of staff and teaching programmes during a period when the college held a pre-eminent position in the teaching of science in England.[7]

Taylor's influence waxed and waned during the period, however, with continued variations in the time and effort which he devoted to the college. In the autumn of 1850, for example, he became seriously ill and endeavoured to resign from the treasurership. The committee, however, prevailed upon him to stay and delegated his functions, for what was to be a period of nearly two years, to a permanent official of the college. The precise nature of Taylor's illness is not clear but it might well have been a mild stroke or heart attack induced by his still hectic pace of life in advancing years. Whatever it was, he did not really begin to recover until the summer of 1851, when he again began to travel and gradually resumed the responsibilities as treasurer. As late as February 1852 he was still obliged to write to the college, apologising for a failure to submit accounts, because of a complaint, 'which prevents me engaging in litterary pursuits'. It was not until the summer of 1852 that he seems finally to have recovered his health and reinstated himself firmly back in office.

Taylor's illness was also to prove a turning point in his business life. Although he continued to maintain a major interest in the London operations of the family business and as late as 1856, at the age of 77, was still occasionally kept from college affairs by engagements in the city, he began a process of gradual retirement, devoting an increasing proportion of his time to college affairs and his other social activities. He began to be widely regarded as the 'grand old man' of science and mining and at the college was accorded the honour of taking the chair at the annual meetings of proprietors during the mid-1850s in place of the absent Lord Brougham. When Taylor again became ill towards the end of 1859 and declined to stand for further re-election the following February, his decision was most reluctantly accepted by the general meeting and they insisted on returning him as an ordinary member of council; a position which continued ill-health forced him to retire from the following year. During his time with the college he had made many donations to the library and on his retirement presented a complete collection of the *Philosophical Transactions* 1825-63 and a copy of the *Hebrew Concordance of the Bible*, the work of his ancestor, John Taylor of Norwich.[8]

IV The British Association for the Advancement of Science

Through friendships made and consolidated during the early days of University College, John Taylor also became involved in the foundation of the British Association for the Advancement of Science, with its similar objectives of revitalising and furthering the development of science and technology in Britain. Although already a Fellow of the Royal Society, to which he was elected in May 1825, and well-known for his scientific interests and contributions, John was overshadowed at the founder meeting of the Association in York in 1832 by a glittering list of eminent scientists such as Murchison, Babbage, Faraday, Greenough, Sedgwick and many others from home and abroad. He accordingly received little mention in reports of that occasion. However, his considerable unassuming organisational talents were to be employed in full in ensuring the success and continued development of the Association. As one of his obituaries declared, 'of the several societies to which he belonged, the British Association for the Advancement of Science has been the most indebted to Mr Taylor's useful co-operation, both scientific and administrative.'[9] It was at John Taylor's London home, 12 Bedford Street, Lincoln's Inn, that the Association's first council meeting was held on 26 June 1832 and they were to enjoy the benefits of his great financial expertise as treasurer for their first thirty years of activity, until his retirement in September 1861. It was a remarkable period of continuous service which kept him in touch with every aspect of the Association's business and carried him into close contact with many other similarly motivated institutions, such as the Society for the Diffusion of Useful Knowledge.[10]

Through the auspices of the British Association, John Taylor was able to make considerable progress in another of his life-long crusades, namely making provision for the regular and systematic collection of certain basic mining data. His experience as a mine agent and speculator had demonstrated time and again the financial, engineering and safety problems which could arise from not having ready access to simple information such as careful plans of underground workings, showing where and how lodes had been exploited in the past; price and production data on a local and national level; and indexes of the relative efficiency of various types of steam engines. He had often endeavoured to calculate and publish certain types of this information himself but what was really needed, he argued, was *systematic and regular collection* to show trends and changes over time and for this he personally had neither the time nor the resources.

A model for the type of information he called for was provided by Thomas Lean's *Engine Reporter* which published monthly data on the performance and efficiency of a sample of pumping and winding engines in Cornwall. This had once provided a medium for a 'race' between mine owners in Cornwall to improve the design and efficiency of their engines, but by the late 1830s interest in this field was flagging. In an attempt to revive activity he persuaded the Association to encourage Lean to publish aspects of his work in *Historical Statement of the Improvements made in the Duty Performed by the Steam Engines in Cornwall* (1839), which became one of the most important nineteenth century commentaries on steam technology.[11]

More importantly, Taylor took steps to ensure that one of the earliest sub-committees established by the British Association was that to promote the collection of mining records. It included many of the most informed authorities of the day. The chairman was Sir Charles Lemon, a mine speculator, MP for Penryn and later West Cornwall, and the leading representative of the Cornish mining interest in Parliament, while the ordinary members included the noted geologists and chemists, Professors Sedgwick, Buckland and Phillips, as well as Henry de la Beche, later first director of the Geological Survey, Mr Sopwith, chief agent for the large Blackett-Beaumont lead mining operations in the North Pennines, and the noted author and industrial commentator, G. R. Porter. Taylor acted as secretary to the committee. Largely through the efforts of this committee in collecting material and making representations to the Chancellor of the Exchequer, the government-financed Museum of Economic Geology was established at the end of 1838 with a special office, set up in 1839, to receive material from the British Association and other sources. The first curator of the Museum was Richard Phillips, a member of the committee, Fellow of the Royal Society, and editor of the *Annals of Philosophy* and the *Philosophical Magazine.* He was one of John Taylor's long-standing friends and had nominated him for election to the Royal Society. They held frequent discussions on analytical chemistry and mineralogy and Phillips often conducted difficult assays and analyses for Taylor on domestic and foreign ores. It was at least partly due to Taylor's influence that he received the appointment.[12]

Initially the museum's terms of reference satisfied only part of Taylor's requirements. It appears to have been limited only to receiving information deposited with it and not to have any regular or systematic collecting function. Accordingly, the next few years saw efforts by Taylor and others to extend its functions. In 1844, for example, he joined with William Bainbridge, a Newcastle barrister and an authority on mining law, in calling for a system of regional mine registration offices to be co-ordinated through the Museum of Economic Geology. Under this arrangement coal and metal mines would be obliged to deposit up-to-date plans in these offices and therefore gradually build up a continuing and expanding record of all underground workings in the country. This, it was hoped, would prove of great practical benefit to current and future generations of mining engineers, provide a basis for the settlement of legal disputes, and reduce the possibilities of serious accidents from unsuspectingly breaking into old flooded workings.[13] Unfortunately there was little immediate support for such a system and it was not until the 1870s that the registration of mine plans became compulsory. Even then inaccuracies in the plans and frequent failures to make deposits made the system notoriously unreliable and to this day disasters, such as that at Lofthouse Colliery in 1973, result as a consequence.

More immediate success was achieved with arrangements for the collection and publication of production and price data. Taylor had himself made periodic estimates of national production during the 1820s, '30s and '40s[14] but this was undertaken on a regular basis from 1848 by the Royal School of Mines and later, from 1854, by Robert Hunt in *The Mineral Statistics of the United Kingdom,* published annually. Figures of ore output were given for each mine and county with national aggregates, equivalent metal production and estimates of average prices. In the early reports there was a continued heavy reliance on returns furnished by John Taylor and Sons, then the country's largest ore producers[15] and one of the few offices with good market intelligence from every major mining district.

REFERENCES

1 In 1811 he developed this interest in local history to write an introduction to a new edition of Risdon's *Survey of Devon.*

2 H. B. Woodward, *The History of the Geological Society of London* (1907), p307.

3 D. P. Alford, *The Abbots of Tavistock* (1891), p319.
4 Letter to W. G. Adams, the Duke of Bedford's agent, 12 November 1815, DRO Bedford Collection, Letters 83.
5 John Taylor's obituary in *Proceedings of the Royal Society of London,* **13** (1864), pVI.
6 University College of London, D.M.S. Watson Library, College Correspondence, Letters from John Taylor.
7 *Ibid.* Out-letters; and S. G. Checkland, *The Rise of Industrial Society in England 1815-1885* (1964), p90.
8 University College London, Minutes of Annual General Meetings of the Proprietors, 1826-63.
9 *Proceedings of the Royal Society* **13** (1863-4), pIV.
10 John also encountered this organisation through his activities at University College, which employed the same full-time secretary. See University College London, D.M.S. Watson Library, Society for the Diffusion of Useful Knowledge, Correspondence.
11 D. B. Barton, *The Cornish Beam Engine* (Truro, 1966), p50.
12 Geological Museum Library, GSM 1, No 68, Letter from John Taylor to Henry de la Beche, 10 Feb 1839, giving reference for Richard Phillips.
13 *Ibid,* GSM 12, Original Letters to the Director General (Guard Book), Nos 44 and 47.
14 A. Ure, *Dictionary of Arts, Manufactures and Mines* (1839) see 'Lead'; *Records of Mining* p173; Sir Charles Lemon, 'The Statistics of the Copper Mines of Cornwall' *Journal of the Statistical Society of London* **I** (1838), p83 and *Memoirs of the Geological Survey of Great Britain* **II** pt 2 (1848), pp703, 709.
15 *MJ,* 14 Aug 1847, p386.

19 Taylor and Martineau's high pressure engine, 1825. This was claimed to be the first successful horizontal high-pressure steam engine, and although produced after John Taylor had left the partnership with his brother Philip it is likely that he was involved in its development.

CHAPTER 6

Technical and Scientific Contributions

Although an active member of virtually every great scientific society of the nineteenth century, John Taylor's *direct* contribution to the advancement of science and technology was limited. Early in life he had invented useful devices such as the ventilation machinery designed for use in driving the Morwellham Tunnel and later widely adopted in mines; the roller ore crusher; and together with his brother Philip, patent methods for producing gas and purifying sugar. Thereafter, however, he contributed little in the sense of complete discoveries or inventions. He may have worked with Philip on the development of high pressure boilers and assisted him in the design of the first practical high-pressure horizontal steam engine, manufactured by Harvey's of Hayle under the superintendence of his friend Arthur Woolf from 1825. However, there were no further specific improvements attributable directly to John Taylor other than a patent for a minor arrangement for economising on the fuel and power of steam engines, issued in 1844.[1] From his early forties most of his active scientific interests were concerned with identifying and stating problems, collecting information on them and encouraging others to find a solution. He enjoyed investigation and summarizing the state of existing knowledge but had neither the time nor, perhaps, the ability to provide his own answers. He was at his best when collaborating with the professional and truly talented engineer, providing informed advice and intellectual encouragement and, more practically, the immense financial resources and opportunities which his business empire placed at his command. The conclusion to one of his many articles in the learned journals might stand as an epitaph to this work: 'My intention was rather to state the facts than to attempt an explanation of what is certainly at present very obscure; but that I may do all in my power to direct attention to the subject, I will venture on a supposition.'[2]

The area of Taylor's main collaborative activity was that of steam pumping engines and pumps; subjects which had real and immediate relevance to his everyday practical affairs. Careful records of the performance and experiments with steam power at his mines were furnished to all serious students of the subject, including John Farey for his monumental *Treatise on the Steam Engine.*[3] Farey later acknowledged in this text that:

> Mr Taylor, in his character of many large mining establishments, by giving encouragement to able practical engineers, has greatly promoted the progress of improvements in steam engines and machinery used in mining, and has extended the Cornish system of engines and pumps into North Wales, from whence it may be expected to spread farther, when its merits are more generally known.[4]

This work first became important in 1818 when Taylor began his long collaboration with Arthur Woolf at Consolidated and several other of his Cornish mines. At this time and into the 1820s most of the steam engines employed in mining were low-pressure machines, sometimes employing two cylinders on the compound system developed by Woolf earlier in the century. The efficiency of these engines in power and fuel consumption was considerably greater than that achieved by the earlier Newcomen and Watt engines, but the continued rate of improvement was slow and many had come to believe that no further major advances could be achieved. Woolf, however, whose very continued employment as an engine designer was threatened by such opinions, was convinced that great improvements in efficiency were obtainable from an alternative system using a single cylinder supplied with high-pressure steam. The steam would be admitted to the cylinder only for the first part of the stroke and then cut-off, the piston continuing to be moved by the natural expansion of the steam trapped in the upper part of the cylinder.

Taylor was the only major mining adventurer to appreciate the possibilities of such a system and to offer the finance and facilities for it to be accurately tested. At Wheal Alfred he and Woolf arranged for the installation and comparative trials of an engine of this new design with a standard compound engine, then regarded as the most efficient type available. The trials began in the autumn of 1824 and continued over the next twelve months. From the carefully kept records, Taylor was able to calculate that in 1825 the new and old engines together had been able to drain the mine for a consumption of 8,053 bushels of coal per month. This compared with 14,390 bushels per month consumed by the old engines alone in 1816 when draining a much smaller and shallower area and represented a saving of 6,337 bushels of coal per month, or 44 per cent of the former fuel consumption. As Farey was later to observe, 'The results of this comparison was considered as decisive in favour of engines with one large cylinder; and, in consequence, the compound engines have since been entirely laid aside in Cornwall'.[5] Farey also noticed that these experiments and others carried by John Taylor and one of his senior captains, William Francis, in 1828, also incontestably proved the law of Boyle and Gay-Lussac that the elasticity (or pressure) retained by steam and air after expansion is inversely proportionate to the volume that it has then acquired.[6] It is tempting to speculate that it was prior knowledge of these theoretical laws, gained through membership of the learned societies, that first prompted Taylor to undertake this experimental work. Unfortunately there in no direct evidence of this.

Following this strategic breakthrough, efficiency figures again began to advance rapidly as the new engines were improved and developed. Interest revived and publications reporting the performance of engines at the major Cornish mines blossomed. By the early 1830s, however, many interested groups, particularly outside Cornwall, began to express incredulity at the performances claimed and the way in which they were calculated. In defence of their accuracy and, more importantly, the whole study of steam engine efficiency and further improvements, Taylor published several papers in the *Philosophical Magazine* and the *Mining Review* drawing on his indisputable first-hand knowledge of the Consolidated, United, and several other mines in the district. As well as reiterating the findings of his experiments at Wheal Alfred, he produced many other specific practical examples of the way in which efficiency had been improved, including the calculation that at the Consolidated Mines in 1830 six engines consuming 11,601 bushels of coal per month could drain an area twice as large and several hundred feet deeper than 30 years previously when four Watt engines had consumed more than another 1,000 bushels per month.[7] John Taylor declared that using the same methods he had carefully studied the duty of the improved engines introduced into his mines in Wales and Mexico and everywhere there had been a marked increase in efficiency. He concluded that,

> Much has been said respecting the motives of those who have published statements on this subject, and therefore I repeat for myself, and other adventurers in the mines of Cornwall, that we have no interest whatever in inducing others to give credit to these statements. We know that we are reaping great advantage from the use of certain improvements, and we might be content to keep this knowledge to ourselves: we are not manufacturers of the engines, but the purchasers of them; we have no monopoly to maintain or any other advantage to enjoy, but what may be shared by every person who may wish to employ similar engines; but we do feel it to be an act of justice to those ingenious men by whose labours we are benefitted, to bear testimony to the success of their efforts; and, as since the time of Boulton and Watt, no one who has improved our engines has reaped pecuniary reward, it is at least fair, that they should have credit for their skill and exertion. We are not the partisans of any individual engineer or engine maker: we avail ourselves of the assistance of many; and the great scale upon which we have to experiment makes the result most interesting to us.[8]

As well as his active participation in the improvement of steam power, Taylor also became interested in the design and operation of the pumps to which that power was applied. He pioneered the introduction of the more efficient plunger pump in place of the traditional lifting pump in his Cornish and Welsh mines and, together with Woolf, suggested detailed improvements to their design. Again he also took a leading role in publicising the advantages of this new system in several publications, including his paper 'On Pumps Used in Mines' in his *Records of Mining.* These activities as innovator and publicist, both here and across the whole spectrum of mining and dressing technology, made him one of the most informed and influential general authorities of his day; a reputation which was reflected in the great respect afforded him by all of those who were knowledgeable of mining affairs. Without ever taking a direct role himself in the initiation or invention of a strategic new technology, Taylor, perhaps more than any other single figure, was responsible for transforming British mining practice from the province of empirically orientated practical engineers into a rigorous scientific discipline, where problems were systematically sought out and solved by scientific methods and the use of the latest scientific knowledge.

Apart from his active involvement in mining technology, John Taylor's main contribution through the learned societies was again principally that of an administrator, innovator and what might be termed a 'pollinator'. The great business acumen which led him constantly to search for practical uses for the latest scientific advances, helped ensure that science remained linked to the requirements of the real world, while his ability to receive structure and communicate ideas between widely different fields of scientific enquiry helped provide the cross-fertilisation which was essential for the continued intellectual vitality of scientific development. Without such men science would become arid and irrelevant to the practical improvement of man's material condition; science would not stimulate technology and would not itself receive the essential practical and financial stimuli which it requires from society.[9] It has been a shortage of such all-round men, capable of 'marrying' science and technology, rather than any particular inadequacy in scientific endeavour and ability that has most contributed to Britain's decline relative to other industrial nations since the last quarter of the nineteenth century. John Taylor's contemporaries recognised and honoured him for these abilities. One obituary noted, for example, that he, 'occupied for many years a prominent position in the scientific world' although he was, 'never strictly a scientific man'.[10] Another wrote more specifically that,

20 A model of Taylor's Cornish pumping engine erected at the United Mines, Gwennap, in 1840. This engine had a steam cylinder of 85in diameter and lifted 550 gallons per minute. *(Science Museum, London).*

> throughout his eminently practical life Mr. Taylor strove constantly to maintain a healthy reciprocity between science and practice; and while availing himself of this scientific knowledge for the improvement of technical processes, he was no less earnest in turning to account his experience and opportunities as a mining engineer and metallurgist for the promotion of the sciences of geology, mineralogy and chemistry.[11]

He strove constantly to interpret a wealth of past empirical experience into general theories for future guidance. This was an objective which had its short-term successes but which he never fully realised. It was an inadequacy which he honestly confessed. Thus he wrote praising Babbage's work,

> When I first knew mining, there was no system. I have done something to make one, others have done much more, but we have all been guided rather by an impulse than by any well-defined rules – I have been a manufacture also to a considerable extent and here again although as you have found there has been considerable skill in making arrangements, there was wanting a code which you have made for us all.[12]

It is clearly unlikely that John could have had either the time or the capacity to converse with and absorb the ideas of all of those who he might have met through his membership and offices in numerous societies, but it is equally certain that his friendships were diverse and extensive. Of their number a few names stand out. From his early and long association with the Geological Society he became well acquainted with many of the leading academic geologists of the day, such as George Greenough, William Buckland, William Fitton, Adam Sedgwick, Roderick Murchison and Charles Lyell and gave them assistance in their efforts to establish the Geological Survey and Museum in 1835. Its first director, Henry de la Beche, also became a close acquaintance, and the support of Buckland, Sedgwick and Fitton, together with Mark Brunel, Charles Babbage, George Rennie, Davies Gilbert and others, all of whom recommended him from good personal knowledge, was responsible for his election to the Royal Society in 1825.[13] Through repeated membership of the council of that society he came into contact with Sir Humphrey Davy, Sir John Herschel, Sir Edward Sabine, Michael Faraday, Alexander von Humboldt, Charles Darwin, Sir George Airey, Justus Liebig, Sir Charles Lemon, Isambard Brunel and a host of others, many of whom became close friends and frequent visitors to his home. Some of these associations were also reinforced by joint membership of the Council of the British Association, where John also worked with the chemists William Harcourt and Charles Daubeny, the engineers Sir William Fairbairn and Sir William Armstrong and the scientist/politicians the Duke of Argyll, the Marquis of Lansdowne and Lord Egerton. In February 1839, Taylor was elected an early member of the Institution of Civil Engineers, which gave him further opportunities to consolidate his associations with Brunel and Armstrong and introduced him to Joseph Whitworth, John Rennie, George Stephenson, William Cubbitt and the railway magnate Mark Huish. He may also have first met William Pole here, for whose book, *A Treatise on the Cornish Pumping Engine* (1844), he contributed much of the background information.[14]

Many of those who were not part of John's own immediate circle were often friends of his brother Philip, who had established an international reputation as a chemist and engineer in the 1820s. Through their frequent meetings and correspondence, John Taylor undoubtedly gathered an improved understanding of the work of such men as James Nasmyth and Charles Macintosh, Philip's scientific partners for a number of years, as well as Maudslay, Stephenson, Wheatstone, Ricardo, Arago, Gay-Lussac and Jean Baptiste-Say. In a similar way John was introduced to the world of music and art by his brother Edward, once a singer at Covent Garden, musical editor of the *Spectator* and Professor of Music, and to the world of publishing, local politics and botanical studies by his brother Richard. Richard had established a printing works in London with his father in 1803 and often worked in conjunction with John, publishing company prospectuses and the proceedings of learned societies, including the *Philosophical Transactions,* of which he was also editor for some years. In 1837 he persuaded John to become a member of the Linnean Society, a group interested in botanical and zoological studies, of which Richard was Assistant Secretary.[15]

Later in life, when unable to travel and attend as many meetings as formerly, John Taylor could be kept in touch with the work of old colleagues and the proceedings of several societies by his two sons, John and Richard, who became members in their own right. For example, both were active in the Royal Cornwall Polytechnic Society, Richard was elected a Fellow of the Geological Society, and John Jnr occasionally attended meetings of the Institution of Civil Engineers. Their eldest sister, Anne, married Philip Worsley, an active partner in Whitbread's brewery and their eldest son re-established John's own early connection with the chemical industry. John's second daughter, Susan, established aristocratic connections for the family when she married the eminent London physician, Edward Rigby, brother of Lady Eastlake.

This pattern of personal and family friendships, reinforced and widened still further through membership of élitist clubs such as the Athenaeum[16] created for John Taylor a complex network of contacts, information and influence reaching into every aspect of early Victorian scientific, economic and social life. It was a fascinating world without social or intellectual barriers: a world in which John met on equal terms the famous and the powerful and entertained such disparate characters as the Brunels, Babbage, and Felix Mendelssohn. It is remarkable and perhaps a profound comment on the character of the man, that John appears seldom to have used this influence for his own social or political aggrandisement. For example, when he obtained a position of some influence on the Parliamentry candidate selection committee for the Finsbury district of London in 1832, he used that influence not to advance any political ambitions of his own but rather to pursuade his close friend and associate Charles Babbage to accept the nomination. At the same time he also had influential connections in the Clerkenwell constituancy and, not withstanding the great pressure of his own business affairs at this time, offered to organise a publicity campaign on Babbage's behalf.[17] This self-effacing stance goes far in explaining why he never received the national honours to which he might have held claim.

Like his own parents, John Taylor appears to have had a quiet confidence in his own position and to have found full satisfaction in his work, interests and family. Such a picture is certainly suggested in a letter of thanks from Felix Mendelssohn following a visit to John and his family at Coed Du in 1824.

> 35 Bury Street, St. James's.
>
> My dear Sir, – At my departure from Coed Du I was not able fully to express to you and to Mrs. Taylor my gratitude for all the kindness you were pleased to show me during the happy time I passed in your house, and I thought to bury in silence those feelings for which I could find but insufficient expression. However, being returned to this town, and recapitulating in my memory every hour I enjoyed since my absence from London, the recollection of your house and family must again and again occur to my mind, and prevails upon me to address you once more on this subject, though I am aware of how little moment my thanks may be to you. For you are accustomed to see everybody around you happy, and indebted to you for their happiness – the peasants, whose barren ground you have changed into a fruitful garden, as well as your happy children; and accordingly you do not want to hear repeated by a stranger the same feelings which *they* may better and more properly express to you. But I myself cannot forbear uttering what so strongly and most heartily I feel: restrained since long within the rules of ceremony, deprived of a familiarity to which I was accustomed before this journey, in short, after having been a stranger amongst strangers, I was received in your house like a friend. Absent for the first time

from home and from my family, I had at least the pleasure of witnessing a happiness which at a former period of my life I shared myself. Those are the enjoyments which I owe to your and Mrs. Taylor's kindness, and I never shall forget them.

If I should be so happy as to meet in my native land with you and your sons, or with any of your friends, I sincerely hope to find an opportunity of proving how much I feel indebted to you.

Believe me, my dear sir,
Yours very truly,
FELIX MENDELSSOHN BARTHOLDY.[18]

There is perhaps no better proof of the sincerity of these remarks than the spirit of *The Rivulet* which Mendelssohn composed during his brief stay.

REFERENCES

1 No 10,201.
2 John Taylor, 'On the Accidents incident to steam Boilers' *Philosophical Magazine,* 2nd series, **I** (1827), p132.
3 BM, Add Mss 37192 f235.
4 J. Farey, *A Treatise on the Steam Engine* (Newton Abbot, 1971) **II,** p165.
5 *Ibid,* p205.
6 *Ibid,* p53.
7 *Ibid,* p248.
8 *Ibid,* p251.
9 J. D. Bernal, *Science and Industry in the Nineteenth Century* (1953).
10 *Mining and Smelting Magazine* **III** (May 1863), p270.
11 *Proceedings of the Royal Society* **13** (1864).
12 BM, Add Mss 37201, f553.
13 Library of the Royal Society, Cert VII, 148.
14 *Proceedings of the Institution of Civil Engineers* **I** (1839) and following.
15 *Dictionary of National Biography;* F. Boase, *Modern English Biography* (Truro, 1901) **III.**
16 John was elected a member of the Athenaeum in 1825. A lithograph from a portrait painted for the club in 1825-6 by Sir Thomas Lawrence is reproduced in this book.
17 BM, Add Mss 37186; 37187; 37188; 37191; 37192; 37193. Correspondence between John Taylor and Charles Babbage concerning politics, The British Association and manuscripts in preparation.
18 P. M. Taylor, *Memoir,* pp20-1.

Conclusion

Born in 1779, John Taylor established his career and reputation during the latter part of the classical Industrial Revolution period and was in the forefront of the rapid technical and scientific progress which underpinned the world industrial supremacy of mid-nineteenth century Victorian England. He played as large a part in bringing fundamental modernisation to metal mining as Josiah Wedgwood, Richard Arkwright and Matthew Boulton achieved in various sectors of manufacturing industry and Robert Bakewell, Thomas Coke and others produced in agriculture. Like most of these great figures, his actual and most important contributions were not as inventor but as innovator, improver, manager and publicist. These were the essential ingredients of success. Like Matthew Boulton he looked principally to others for the genius of invention and then combined it with his own equally rare and valuable gifts of technical understanding and business ability to create something that was truly practical and financially viable. Within his industry he had no serious rival. No one pursued technological efficiency and financial economy so knowledgeably, boldly and steadfastly; no one operated on such a large and diverse national and international scale; no one had the same record of success, not even the great Williams family of Cornwall, who several times failed where he succeeded.

Outside of business, in the field of science and learning, his contribution was almost as impressive. Not only did he prove an invaluable servant to several of the most influential learned societies, applying his great administrative and financial talents to ensure their proper functioning and development, but he also took a major role in 'marrying' the divergent worlds of science and technology; providing the cross fertilization of ideas, aims and objectives which was so essential for future progress but which was so generally lacking in the over-confident atmosphere of mid-Victorian England.

We must beware of too much eulogy, however, as John Taylor's record was not unblemished by errors of judgment and even monumental failures. The key to his general success was always to show great tenacity of working and investment in a project until a situation of profitability was finally achieved. It was this unflinching faith in the long-term accuracy of his judgment that enabled him and his financial backers to succeed where many others had previously tried and prematurely abandoned a mine. It could even be suggested that in a country still rich in accessible minerals and with a rapidly rising demand for metals consequent on the contemporary process of industrialisation, that such a formula was in most cases a sure guide to success. However, on the few occasions when his judgment was clouded and proved fundamentally wrong, it was an equally sure formula for total failure and unparalleled financial losses. The Mexican enterprise stands and a pre-eminent example. After permitting himself, against his considered judgment, to be caught up in the fever of speculation in South American mines, he found himself trapped in a situation where his entire reputation and the trust of his supporters was dependent on his continuing to pour money into an apparently bottomless pit of disappointment. The losses of the Real del Monte company continued to escalate inexorably long after most other enterprises in the area had been wisely abandoned and to the last he was obliged to argue for still further effort. As it finally turned out this may have been rewarded but at the time few people, including Taylor, could have had much confidence left. A similar situation also nearly occurred at his Derbyshire mines but he appears to have learnt from his earlier experience and managed to extract himself and the company before such disastrous consequences befell them.

Nevertheless, the record is an impressive one. Starting in an unfamiliar industry at an early age with little more than the initial help of a few influential friends, a good education, and a strict but unbigoted religious upbringing which had given him confidence in himself and wide-ranging interests in the affairs of the world, he quickly created the foundations of a business empire which came to rival in size and wealth the largest in England. When he died, on Easter Sunday, 5 April 1863, after a long illness which had left him feeble and almost completely incapacitated, he bequeathed a personal estate of nearly £40,000 and a family business which continued to grow and prosper for more than another century.[1]

Appendix

Bibliography of Publications by John Taylor[2]

'Sketch of the history of mining in Devon and Cornwall', *Phil Mag* **V** (1799), pp357-65.

'Sketch of the general history of mining', *Phil Mag* **XV** (1803), pp164-67.

'On blasting rocks and tamping', *Phil Mag* **XXX** (1808), pp97-101.

---

1 John Taylor's will, Somerset House, 1863, Vol 10 (466).

2 Many of the articles were published in similar form in different periodicals.

'Account of the construction of a machine for conveying fresh air underground', *Trans Soc Arts* **XXVI** (1810).

'Communication on water-pressure engines' *Phil Mag* **XXXVII** (1811), pp129-31.

'Method of ventilating mines or hospitals by extracting foul air from them', *Phil Mag* **XXXVIII** (1811), pp120-7.

'Remarks on the present state of the county of Devon, introductory to the new edition of Risdon's survey' in T. Risdon, *The Chorographical Description or Survey of the County of Devon* (1811), pp i-xxxvi.

'On the ventilation of coal mines', *Ann Phil* **III** (1814), pp196-202.

'On the economy of the mines of Cornwall and Devon', *Trans Geol Soc* **12** (1814), pp309-27. pp309-27.

'Description of the tunnel of the Tavistock Canal, through Morwel Down, in the county of Devon', *Trans Geol Soc* **IV** (1817), pp146-55.

'On some remarkable appearances in coke', *Trans Geol Soc* **IV** (1817), pp448-9.

'On the construction of fire places in boilers', *Ann Phil* **XII** (1818), pp51-2.

Letter to the Editor on the subject of colliery accidents, *The Times* 6 Nov 1821.

'On the smelting of tin ores in Cornwall and Devonshire', *Trans Geol Soc* **V** (1821), pp358-68.

'Observations on Mr Schoolcraft's "Account of the native copper on the southern shore of Lake Superior, with historical citations and miscellaneous remarks" ', *Ann Phil*, New Ser, **III** (1822), pp56-9.

'A course of lectures on metallurgy delivered at the London Institution', *Phil Mag* **LXI** (1823), pp283-91; 368-74; 448-58.

Edited with introduction, *Selections from the works of the Baron de Humboldt, Relating to the Climate, Inhabitants, productions, and Mines of Mexico* (1824).

'Notice respecting a copper mine at Cally in Kircudbrightshire', *Trans Geol Soc*, 2nd Ser, **I** (1824), pp164-5.

Letter on the Mexican mines, *The Times* 22 Nov 1824, p3.

Statements Respecting the Profits of Mining in England Considered in Relation to the Prospects of Mining in Mexico (1825).

'Prospectus of a School of Mines in Cornwall', *Phil Mag* **LXVI** (1825), pp137-50.

'On the accidents incident to steam boilers', *Phil Mag*, 2nd Ser, **I** (1827), pp126-33.

'On rain gauges', *Phil Mag*, 2nd Ser, **II** (1827), pp406-8.

Records of Mining (1829).

'Improvements in the management of mines', *Phil Mag*, 2nd Ser, **VI** (1829), pp383-92.

'On the duty of steam engines in Cornwall', *Phil Mag*, 2nd Ser, **VII** (1830), pp424-31.

'On the duty on steam engines. Revised from the original in the *Records of Mining* and a subsequent paper in the *Philosophical Magazine* and continued to the present time', *English's Quarterly Mining Review* **II** (1831), pp33-59.

'On the collection and arrangement of veinstones and the importance of an accurate examination of their connection with the rocks in which they occur', *Reports and Transactions of the British Association* (1831-2), pp578-80.

'Notices of some improvements in dressing ores', *Quarterly Mining Review* **II** (1831), pp82-7.

'Sur le travail utile des machines a feu en Angleterre et Notamment dans le comte de cornouailles', *Annales des Mines* **II** (1832), pp51-72.

'Description d'un appariel pour le distillation des amalgames', *Annales des Mines* **II** (1832), pp199-202.

'Report on the state of knowledge respecting mineral veins', *Reports and Transactions of the British Association* (1833), pp1-25.

'Account of the depth of mines', *Reports and Transactions of the British Association* (1833), pp427-30.

'On the monthly reports of the duty of steam engines employed in draining the mines of Cornwall', *Reports and Transactions of the British Association* (1835), p108.

'On a new rotative steam engine', *Phil Mag,* 3rd Ser, **VII** (1835), pp369-70.

'On the history of the rotary steam engines working expansively', *Phil Mag,* 3rd Ser, **VIII** (1836), pp136-8.

'On the improvements in Mining', *Quarterly Mining Review* **VIII** (1836), pp353-61.

'On the comparative value of the Mineral Productions of Great Britain and the rest of Europe', *Reports and Transactions of the British Association,* (1836), pp144-6.

'On the duty of steam engines', *Phil Mag,* 3rd Ser, **VIII** (1836), p67.

'On Mining Statistics', *Quarterly Mining Review* **IX** (1837), pp30-9.

'On peroxide of manganese containing silver, from Mexico', *Phil Mag,* 3rd Ser, **X** (1837), pp279-80.

'Observations on the strata penetrated in sinking a well at Diss in Norfolk', *Trans Geol Soc,* 2nd Ser, **V** (1837), pp137-8.

'On the Economy of Mining', *Quarterly Mining Review* **X** (1837), pp261-72.

Address to a meeting at the 5th annual exhibition at the Polytechnic Hall, Falmouth, 11th-13th Oct 1837, *Report of the Royal Cornwall Polytechnic Society* (1837), pp25-7.

'A note to accompany specimens of quicksilver ore from the mine San Onofre, near the town of El Doctor, Mexico', *Trans Geol Soc,* 2nd Ser, **VI** (1838), p686.

Reply to 'Observations on the Statement of the Committee of the (Gwennap) Consolidated Mines Adventurer etc (Printed for Private Circulation, 1838) BM 08227 ee, 45(2).

'On water power at Wheal Friendship', *Reports and Transactions of the British Association* (1841), p106.

'On the Machines for Raising and Lowering Miners and on the Use of the Iron Wire Rope in the Mining District of the Hartz', *Report of the Royal Cornwall Polytechnic Society* (1842), pp65-9.

'Comments on Discussion about railway cuttings', *Proceedings of the Institute of Civil Engineers* **III** (1843).

'Description of a Machine for raising and lowering miners', *Proceedings of the Institute of Civil Engineers* **III** (1843), pp193-5.

'Notice on a steam engine at Haarlem', *Reports and Transactions of the British Association* (1843), p100.

'A. Rous's Steam engine indicator', *Reports and Transactions of the British Association* (1843), p101.

'Comments on discussion of flow of air in pipes', *Proceedings of the Institute of Civil Engineers* **IV** (1845), pp280-1.

INDEX